AF768382

Turfgrass Weed Control for Professionals, 2024

 Catalog-in-Publication Data on file at the Library of Congress.

978-1-62671-229-4 (paperback)
978-1-62671-215-7 (epdf)

Cover image: Dandelion (*Taraxacum officinale*) and wild violet (*Viola sororia*). Photo by Aaron Patton.

Turfgrass Weed Control for Professionals, 2024

Aaron J. Patton
Matthew T. Elmore

Purdue University Press
West Lafayette, Indiana

Contents

Extension Resources

State extension services provide information and publications about turf, landscapes, trees, crops, vegetables, fruit, soil, pests, and wildlife as well as nutrition, family finances, and many other topics.

To order or download the extension publications from your state extension service, see *Additional Resources*, page 123.

Disclaimer

Some of the information presented in this guide, especially pesticide recommendations, is state-specific. Readers should check with their own cooperative extension services for state-specific information. See page 58 for a list of herbicides commonly used by US turf managers that are not registered for use in the northern United States.

Reference in this publication to any specific commercial product, process, or service, or the use of any trade, firm, or corporation name, is for general informational purposes only and does not constitute an endorsement, recommendation, or certification of any kind by the North Central Turfgrass and Environment working group or their institutions. Individuals using such products assume responsibility for their use in accordance with current directions of the manufacturer.

Abbreviatons Used in This Book

A	acre
ae	acid equivalent
ai	active ingredient
AMS	ammonium sulfate
COC	crop oil concentrate
fl oz	fluid ounce/ ounces
ft^2	square foot/feet
gal	gallon/gallons
lb	pound/pounds
MSO	methylated seed oil
NIS	nonionic surfactant
oz	ounce/ounces
pt	pint/pints
qt	quart/quarts
UAN	urea ammonium nitrate
v/v	volume per volume

Poster for Indentifying Turfgrass Weeds

An important first step to weed management is properly identifying these plants. This poster identifies the most common weeds found in cool-season turf. The poster format can be utilized in the breakroom, chemical room, maintenance shop, or equipment trailer for employee education.

A companion resource for Turfgrass Weed Control for Professionals (TURF-100), sixty weeds are covered in the Identifying Turfgrass Weeds poster (Purdue Extension publication TURF-101). The poster is available from the Education Store—https://edustore.purdue.edu/.

Acknowledgments

The authors wish to thank the Midwest Regional Turf Foundation (MRTF) for their support of turf research and education at Purdue University. This work was also supported by Hatch project 1019791 from the USDA National Institute of Food and Agriculture. Their support aids in the development of the research-based information summarized in this publication.

MRTF
Midwest Regional Turf Foundation

Introduction

Turfgrass management is a highly skilled, technical profession. The need to maintain quality turf areas while protecting the environment dictates that turf professionals be equipped with the latest information about the use of pesticides, fertilizers, and other products. This guide supplies turf herbicide and plant growth regulator information that turfgrass professionals can use to develop effective weed control programs for golf courses, athletic fields, sod farms, residential and commercial lawns, and other turfgrass areas.

Herbicides are useful tools for controlling plants that are poisonous or allergenic or that obstruct traffic line-of-sight in turf. Herbicides also improve turf health by decreasing plant competition for light, water, and nutrients; help desirable turf establish; increase property values and aesthetics; and reduce the labor associated with hand-pulling weeds. Herbicide applications also reduce how frequently turf needs mown (including labor and mower emissions) by controlling weeds that grow more quickly than turf. Herbicides also control weeds that can potentially cause physical injury or discomfort in parks or athletic fields (such as lawn burweed, sandbur, and puncturevine).

This publication, provides specific cultural and chemical recommendations that are based on research experiments at Purdue University and other universities and on information from manufacturer labels. Unfortunately, it is impossible to test every product under every possible set of conditions. Because environmental conditions and application methods may vary widely, a particular herbicide's performance will not always conform to the turf safety and weed control standards indicated in this publication. Also, be aware that herbicide performance varies seasonally because environmental conditions (temperature, rainfall, humidity, and so on) vary, and a large number of changing variables will affect how weeds grow and how herbicides work to control weeds once applied.

Turfgrass Culture

The best defense against weeds is to culture healthy turf. Proper cultural practices can reduce weed populations by 70 percent or more, so herbicides should not be a substitute for a conscientious cultural program. When you choose the right turfgrass species and cultivar for the site and follow proven fertilization, mowing, and irrigation practices, weeds will be less competitive in turf.

Prevention is the key for turf managers who want to avoid introducing weeds into an uninfested area. For example, cleaning weed seed off equipment can prevent the spread of plants. Implementing proper cultural practices is the cornerstone of effective integrated pest management for turfgrass weed control.

Before implementing any weed control program, first determine why the turf is thin and why weeds are invading (see *Indicator Weeds*, page 4). Correct the underlying causes of unhealthy turf before implementing a program using herbicides.

This publication discusses several turfgrass species. The cool-season turfgrasses discussed include annual bluegrass (*Poa annua*), creeping bentgrass (*Agrostis stolonifera*), fine fescue

Kentucky bluegrass

Tall fescue

Creeping bentgrass

Buffalograss

Perennial ryegrass

Fine fescue (Chewings fescue shown)

Bermudagrass

Zoysiagrass

(*Festuca* spp.), Kentucky bluegrass (*Poa pratensis*), perennial ryegrass (*Lolium perenne*), and tall fescue (*Schedonorus arundinaceus*). Each of these cool-season turfgrasses can be found throughout the Midwest and Northeast. Five fescue species are commonly used as turf, but we typically simplify them into two groups: fine fescues and tall fescue.

Fine fescues include strong creeping red fescue (*Festuca rubra* ssp. *rubra*), slender creeping red fescue, (*Festuca rubra* ssp. *littoralis*), hard fescue (*Festuca brevipila*), and Chewings fescue (*Festuca rubra* ssp. *commutata*). These fine fescues have a narrow or slender leaf blade and are all very similar to one another in appearance and herbicide tolerance. For these reasons, this publication will refer to the fine fescue species as a group.

Tall fescue—sometimes called turf-type tall fescue—is a wider bladed turfgrass species used for lawns and as forage, depending on the cultivar.

Each of these cool-season turfgrasses can be found throughout the Midwest.

Among the warm-season grasses grown in the Midwest, common bermudagrass (*Cynodon dactylon*), buffalograss (*Buchloë dactyloides*), and zoysiagrass (*Zoysia* spp.) are discussed in this publication. These warm-season turfgrasses are primarily grown in southern regions of the Midwest and mid-Atlantic states in addition to the Southeast and Southwest.

In some cases, these turfgrasses can be defined as weeds, while in other cases they are the preferred ground cover.

Weed Types

A weed can be simply defined as any plant that is objectionable or interferes with the activities or welfare of people.

An important first step to weed management is to properly identify the weeds you want to control. It is important to identify the type of weed because that will determine your control options. Seven common weed types in turf are:

1. Grasses
2. Broadleaves
3. Sedges
4. Rushes
5. Lilies
6. Spiderworts
7. Other weeds

Weed identification should begin with classifying weeds by type. The three most common weed types are grasses, broadleaves, and sedges.

Grasses

Grasses are monocotyledonous plants (monocots), which means they have only one seed cotyledon (leaf) present when seedlings emerge from the soil. Grasses have joints (nodes) and hollow, rounded stems. The true leaves of grasses (as opposed to seed or cotyledon leaves) have parallel veins and are several times longer than they are wide. Crabgrass, goosegrass, quackgrass, and annual bluegrass are common grassy weeds found in turf.

Broadleaves

Broadleaf weeds are dicotyledonous plants (dicots), which means they have two cotyledons when seedlings emerge and have net-like veins in their true leaves. Broadleaves often have colorful flowers compared to the inconspicuous flowers found on grasses. Common chickweed, henbit, broadleaf plantain, white clover, and dandelion are typical broadleaf weeds.

Sedges

Sedges are monocots that have solid, triangular stems (in most species) that bear leaves extending in three directions (3-ranked). Sedges lack ligules and auricles and they each have a leaf sheath that is continuous around the stem. Yellow nutsedge and false-green kyllinga are common sedge weeds.

Rushes

Rushes are monocots that have have round, solid stems and favor moist habitats. Path rush is a common rush weed, which can be found near golf cart paths, sports fields, and other compacted areas.

Lilies

Lilies are monocots that have linear leaves, mostly with parallel veins. Some lilies have bulbs while others have rhizomes. Wild garlic, wild onion, and star-of-Bethlehem, are members of the lily family.

Spiderworts

Spiderworts are monocots that have small, three-petaled flowers, fleshy stems, and leaves with parallel veins. Doveweed (a southern U.S weed) and Asiatic dayflower are examples of spiderwort weeds.

Other Weeds

Not all turfgrass weeds fall into these six categories. Some weeds (such as scouringrush) fall into a unique group. It is not a monocot or a dicot but instead a horsetail. Moss is another example of a weed that is not a monocot or a dicot but instead a bryophyte.

See *Weed Identification*, pages 5–35 for color pictures and more information about the most common turf weeds in the northern half of the United States and southern Canada.

Weed Life Cycles

Weeds may be further divided by their life cycles: into annuals, biennials, and perennials.

Annuals

Annuals germinate from seed, grow, mature, and die in less than 12 months. Annuals may be further classified as winter or summer annuals.

Winter annuals germinate in the fall, overwinter as plants, mature in the spring, flower, set seed and then die during the summer. Annual bluegrass is our most common winter annual grassy weed in turf. Henbit and common chickweed are examples of winter annual broadleaves.

Summer annuals germinate in the spring, grow actively during the summer, flower, set seed in late summer and die in the fall. Crabgrass and goosegrass are summer annual grasses. Prostrate spurge is an example of a summer annual broadleaf.

Biennials

Biennials reproduce from seed and complete their life cycle in two years. Biennials form rosettes and store food in their fleshy roots the first year, and then flower the second year. There are few biennial weeds in turf: wild carrot, musk thistle, and bull thistle are the most common biennial turf weeds.

Perennials

Perennial weeds live more than two years. They may reproduce from seed or from vegetative structures such as roots, rhizomes, stolons, tubers, corms, or bulbs. This ability to reproduce vegetatively makes perennials more difficult to control. Some perennials (such as dandelion, curly dock, and wild garlic) actively grow during cool weather, while others (such as field paspalum and yellow nutsedge) grow rapidly during the summer.

Perennials may be further classified as simple perennials or creeping perennials.

Simple perennials (such as curly dock and dandelion) overwinter by means of a vegetative structure (such as a perennial root with a crown), but reproduce almost entirely by seed.

Creeping perennials can both overwinter and produce new, independent plants from vegetative reproductive structures. Vegetative reproductive structures include stolons (creeping bentgrass), rhizomes (quackgrass), tubers (nutsedge), corms (spring beauty), and bulbs (wild garlic). Most perennials can also reproduce from seeds.

Developing a Weed Control Program

There are several important things to consider when you develop a weed control program:

1. Know what kind(s) of turfgrass you have and the total area of each type.
2. Identify the problem weeds and note what time of the year they occur.
3. Determine why the weeds invaded the turf area and correct the conditions or cultural practices that caused the problem (see *Indicator Weeds*, below).
4. When herbicides are needed:
   - Select a chemical that is effective on the weeds and labeled for use on the turfgrass species you are treating.
   - Follow all label directions even if they conflict with this publication—the label is the law.
   - Apply the herbicide at the correct time (time of year and weed life cycle) and rate.
   - Apply the herbicide uniformly over the turf area without skips or overlapping.
   - Repeat the herbicide application at the recommended interval when specified on the label.
5. Follow a sound turf management program in conjunction with the weed control program. An integrated approach—which includes cultural control by enhancing turfgrass competition, using mechanical weed control, and using chemical weed control methods—will be the most successful weed control program.

Indicator Weeds

Indicator weeds are weed species that may help indicate other cultural problems. For example, goosegrass commonly occurs in areas with compacted soils. Therefore, finding goosegrass may be a sign that you need to cultivate the soil to reduce compaction.

When you see a weed, it is a good idea to assess the growing conditions. And remember: although it can be helpful to view weeds as indicators of cultural problems, the presence of a weed does not always indicate a cultural problem.

The *Common Indicator Weeds* table (below) shows some common cultural problems and the weeds that can indicate those problems.

Common Indicator Weeds

Cultural Problem	Indicator Weeds
acidic soils (low soil pH)	red sorrel
alkaline soils (high soil pH)	broadleaf plantain, wild carrot
compacted soils	annual bluegrass, common bermudagrass, common chickweed, goosegrass, mouse-ear chickweed, path rush, pineapple weed, prostrate knotweed, prostrate spurge, puncturevine, windmillgrass
dry soils	field bindweed, red sorrel, sandbur, windmillgrass
dry, low fertility soils	broomsedge, birdsfoot trefoil, black medic, buckhorn plantain, chicory, crownvetch, hop clover, lespedeza, prostrate spurge, sandbur, wild carrot, yarrow, yellow woodsorrel
flooded soils	barnyardgrass, prostrate knotweed, yellow nutsedge
high or infrequent mowing	birdsfoot trefoil, buckhorn plantain, chicory, common mullein, foxtails, hawkweed, johnsongrass, purpletop, red clover, thistles, wild carrot
high-nitrogen fertility soils	annual bluegrass, common bermudagrass, creeping bentgrass, mallow, purslane
high-potassium fertility soils	dandelion
low fertility soils	broadleaf plantain, buckhorn plantain, crownvetch, hop clover, lespedeza, red clover, red sorrel, smooth brome, timothy, vetch, white clover
moist, fertile soils	creeping bentgrass, curly dock, henbit
moist or poorly drained soils	annual bluegrass, annual sedge, barnyardgrass, common chickweed, crabgrass, creeping bentgrass, ground ivy, kyllinga, ladysthumb, moss, mouse-ear chickweed, Pennsylvania smartweed, rough bluegrass, shining flat sedge, speedwells, violets, yellow nutsedge
mowing too low	annual bluegrass, birdseye pearlwort, common bermudagrass, crabgrass, creeping bentgrass, moss, white clover
new seedlings, summer	barnyardgrass, carpetweed, crabgrass, goosegrass, prostrate pigweed, purslane, fall panicum, witchgrass, yellow foxtail
new seedlings, winter	annual bluegrass, annual ryegrass, corn speedwell, field pennycress, henbit, purple deadnettle, purslane speedwell, shepherd's purse, speedwell purslane, wheat
sandy soils	puncturevine, sandbur
shade	annual bluegrass, bulbous bluegrass, Carolina geranium, common chickweed, common plantain, dovefoot geranium, ground ivy, healall, Indian mock-strawberry, lesser celandine, moss, nimblewill, path rush, rough bluegrass, spring beauty, star-of-Bethlehem, violets, woodland sedge, yellow woodsorrel

Weed Identification

How to Use This Section

By definition, a weed is any plant that is objectionable or interferes with the activities or welfare of people. As such, plants covered in this publication may be objectionable in some scenarios and desirable in others. Plants covered in this guide include those plants frequently submitted to Extension for identification or those frequently identified in turf by the authors.

Each entry includes the parts below.

- Plants are arranged alphabetically by family name. These names end in the suffix -aceae. The family names are only listed for the first weed in that family.
- Within each family, plants are arranged alphabetically by their scientific name. Scientific names are binomial (two names) and are written in Latin. They are a universal and used in every country. The general (or genus) name is the first name and it is capitalized. The specific (species) name is the second name given. Together, they make up the plants name and are useful when searching for additional written information on these plants. Scientific names are always *italicized*.
- The **common name** of the plant is listed next. If applicable, we also provide any **other common name**.
- Next, we describe the plant's life cycle (perennial, biennial, summer annual, winter annual) and type (grass, broadleaf, sedge, rush, lily, etc.).
- Two to three bullets follow that provide information about identifying the plant.

Habitat: Describes the environment in which the plant is often found. Use this knowledge to develop cultural control tactics.
Control: Provides guidance about cultural and/or chemical control strategies for the weed.
Similar Species: Lists the names of other plant species with a close resemblance (lookalikes).

Contents

Amaranthaceae (Amaranth Family)

Amaranthus blitoides—Prostrate Pigweed

Other Common Name: mat amaranth

- Summer annual broadleaf
- Prostrate, smooth, fleshy, reddish-purple stems
- Small green flowers
- Leaves are small with a distinct notch at tip

Habitat: Common in disturbed areas and summer seedings
Control: Most broadleaf herbicides are effective
Similar Species: common purslane (page 31), other pigweeds

Apiaceae (Carrot Family)

Daucus carota—Wild Carrot

Other Common Name: Queen Anne's lace

- Biennial broadleaf
- Growth is prostrate in the first year; leaves are pinnately compound
- Hairy stems give rise to white flowers in an umbel
- Mature flower clusters resemble a bird's nest

Habitat: Low-maintenance, unirrigated turf
Control: Treat in the first year; control with 2,4-D is variable.
Similar Species: yarrow (page 6), poison hemlock (*Conium maculatum*)

Asteraceae (Aster Family)

Achillea millefolium—Yarrow

- Perennial broadleaf
- Finely divided, lanceolate leaves
- Stems and leaves have fine hairs
- Flat-topped white or yellow flowers

Habitat: Low maintenance, unirrigated turf. Escapes from landscape beds.
Control: Most broadleaf herbicides are effective
Similar Species: *Chamomile* spp., pineapple weed (page 7), wild carrot (page 6)

Ambrosia artemisiifolia—Common Ragweed

- Summer annual broadleaf
- Leaves are twice compounded (pinnatifid)
- Stems are erect and branched with hairs
- Flowers are green and inconspicuous

Habitat: Disturbed sites or next to agricultural fields
Control: Most broadleaf herbicides are effective
Similar Species: giant ragweed (*Ambrosia trifida*)

Bellis perennis—English Daisy

Other Common Names: English lawndaisy, lawndaisy

- Perennial broadleaf
- Elliptical or round leaves without stems
- Flowers are daisy-like and about 1-inch in diameter

Habitat: Low-maintenance turf and meadows

Control: Penoxsulam is the most effective herbicide

Similar Species: asters

Carduus nutans—Musk Thistle

Other Common Name: nodding plumeless thistle

- Biennial broadleaf
- Rosette growth habit in first year
- Stems are erect and flower the second year; flowers are showy and redish purple
- Leaves are smooth with a whitish midrib and spine-tipped leaf margins

Habitat: Thin turf, roadsides, pastures, etc.

Control: See *Thistles*, page 54

Similar Species: bull thistle (page 8), Canada thistle (page 8)

Chamomilla suaveolens (Matricaria discoidea)—Pineapple Weed

Other Common Name: disc mayweed

- Winter or summer annual broadleaf
- Rosette growth habit with finely divided, pinnately compound leaves
- Yellow flowers have a pineapple-like scent

Habitat: Often found in thin turf, compacted soils

Control: Increase turf density; most broadleaf herbicides are effective.

Similar Species: *Chamomile* spp., wild carrot (page 6), yarrow (page 6)

Cichorium intybus—Chicory

Other Common Name: nodding plumeless thistle

- Perennial broadleaf
- Leaves are toothed or pinnatifid, hairy, and resemble dandelion leaves
- Rosette growth habit with erect flowering stems
- Bright blue flowers produced from June to October

Habitat: Thin turf, roadsides, pastures, etc.

Control: Increase turf density; most broadleaf herbicides are effective.

Similar Species: dandelion (page 10)

Cirsium arvense—Canada Thistle

- Perennial broadleaf
- Leaves with smooth surfaces have spiny margins and flowers are pink to purple
- They aggressively spread from rhizomes, which results in large colonies
- Initially establish from windblown seed

Habitat: Thin turf, location near seed source

Control: See *Thistles*, page 54

Similar Species: bull thistle (page 8), musk thistle (page 7)

Cirsium vulgare—Bull Thistle

- Biennial broadleaf
- Rosette growth habit in first year
- Stems are erect with reddish-purple flowers the second year
- Leaves are hairy with spine-tipped leaf margins

Habitat: Establishes in a wide range of conditions

Control: See *Thistles*, page 54

Similar Species: Canada thistle (page 8), musk thistle (page 7)

Conyza canadensis—Horseweed

Other Common Names: Canada fleabane, marestail

- Winter annual broadleaf
- Rosette growth habit with hairy leaves
- Stems elongate as the plant matures with small leaves
- Numerous, small white flowers; disperses seed by wind

Habitat: Common in areas near a seed source

Control: Treat when small; control with 2,4-D is variable.

Similar Species: annual fleabane (*Erigeron annuus*), Virginia pepperweed (page 11)

Hieracium caespitosum (Hieracium pratense)—Yellow Hawkweed

Other Common Name: meadow hawkweed

- Perennial broadleaf
- Rosette growth habit with narrowly elliptic, hairy leaves
- Produces flower heads in summer after dandelion
- Spreads by seed, stolons, and rhizomes

Habitat: Thin turf, roadsides, pastures, etc.

Control: Most broadleaf herbicides are effective

Similar Species: cat's-ear dandelion (*Hypochaeris radicata*), dandelion, orange hawkweed (page 9)

Hieracium aurantiacum—Orange Hawkweed

- Perennial broadleaf
- Rosette growth habit with narrowly elliptic, hairy leaves
- Characteristics identical to yellow hawkweed, but flowers are bright orange
- Spreads by seed, stolons, and rhizomes

Habitat: Thin turf, roadsides, pastures, etc.

Control: Most broadleaf herbicides are effective

Similar Species: cat's-ear dandelion (*Hypochaeris radicata*), yellow hawkweed (page 8)

Senecio vulgaris—Common Groundsel

Other Common Name: old-man-in-the-spring

- Winter annual broadleaf
- Erect growth habit with deeply lobed leaves
- Erect stems produce yellow flowers; seeds resemble dandelion seeds

Habitat: Bare, nutrient-rich soils

Control: Cultivation or broadleaf herbicides are effective

Similar Species: sowthistles

Sonchus arvensis—Perennial Sowthistle

Other Common Name: field sow thistle

- Perennial broadleaf
- Prickly-toothed leaves and erect stems
- Rhizomatous growth habit
- Yellow flowers and seeds reminiscent of dandelion

Habitat: Roadsides, pastures, etc.

Control: Most broadleaf herbicides are effective

Similar Species: Prickly lettuce (*Lactuca serriola*), other sowthistles

Sonchus oleraceus—Annual Sowthistle

Other Common Name: common sowthistle

- Winter annual broadleaf
- Prickly-toothed leaves and erect stems
- Reproduces by wind-blown seed
- Yellow flowers; seeds resemble dandelion seeds

Habitat: Bare, nutrient-rich soils

Control: Most broadleaf herbicides are effective

Similar Species: spiny sowthistle (*Sonchus asper*), perennial sowthistle (page 9), prickly lettuce (*Lactuca serriola*)

Taraxacum officinale—Dandelion

Other Common Name: common dandelion

- Perennial broadleaf
- Leaves have distinct wavy margins that form into the irregular toothed appearance that makes the plant easy to identify
- Bright yellow flowers transform into a globe-like, grayish-white seedhead

Habitat: Establishes in a wide range of conditions

Control: Several broadleaf herbicides are effective, especially 2,4-D

Similar Species: cat's-ear dandelion (*Hypochaeeris radicata*), chicory (page 7)

Brassicaceae (Mustard Family)

Barbarea vulgaris—Yellow Rocket

Other Common Name: winter cress

- Winter annual or biennial broadleaf
- Rosette growth habit with deep-green, glossy foliage
- Yellow flowers are produced on stems in the second year of growth

Habitat: Thin turf, roadsides, pastures, etc.

Control: Most broadleaf herbicides are effective

Similar Species: Other members of the mustard family

Capsella bursa-pastoris—Shepherd's Purse

- Winter annual broadleaf
- Rosette growth habit with toothed leaves
- Small white flowers are produced on an elongated stem
- Distinctive, heart-shaped, flat seed pod

Habitat: Thin turf, fall seedings

Control: Most broadleaf herbicides are effective

Similar Species: Virginia pepperweed (page 11)

Cardamine parviflora—Small-Flowered Bittercress

Other Common Name: sand bittercress

- Winter annual broadleaf
- Rosette growth habit with pinnate leaves
- Small white flowers are produced on an elongated stem
- Distinctive, long seed pod that explodes to distribute seeds

Habitat: Thin turf, fall seedings, landscape beds

Control: Most broadleaf herbicides are effective

Similar Species: Other bittercresses (*Cardamine* spp.)

Lepidium virginicum—Virginia Pepperweed

Other Common Name: least pepperwort

- Winter annual broadleaf
- Rosette with finely toothed leaves
- Oblong fruit (pods) with a noticeable midrib

Habitat: Thin turf, fall seedings

Control: Most broadleaf herbicides are effective

Similar Species: shepherd's purse (page 10)

Bryaceae (Moss Family)

Bryum argenteum—Silvery Thread Moss

Other Common Name: moss

- Perennial moss
- Very small leaves form dense mats
- Spreads vegetatively and by spores

Habitat: Closely mown, shady, moist sites

Control: See *Moss*, page 52.

Similar Species: birdseye pearlwort (page 11)

Caryophyllaceae (Pink Family)

Cerastium fontanum ssp. *vulgare*—Mouse-Ear Chickweed

Other Common Name: big chickweed

- Perennial broadleaf
- Small, hairy leaves and stems
- Prostrate stems form dense mats
- Five-petaled with white lobed petals

Habitat: Mown turf

Control: Most broadleaf herbicides are effective

Similar Species: common chickweed (page 12), sticky chickweed (*Cerastium glomeratum*)

Sagina procumbens—Birdseye Pearlwort

Other Common Name: birdeye pearlwort

- Perennial broadleaf
- A very small stoloniferous, mat-forming plant
- Small leaves appear whorled
- Flowers and seeds are tiny

Habitat: Found in golf course putting greens and other close mown turf, sidewalk cracks, nurseries

Control: Most broadleaf herbicides should be effective

Similar Species: silvery thread moss (page 11)

Stellaria media—Common Chickweed

- Winter annual broadleaf
- Opposite, elliptic to egg-shaped leaves
- Stems lie prostrate and can root at the nodes
- Five petals with deep lobes give the appearance of 10 petals

Habitat: Shaded, moist sites; thin turf, fall seedings, landscape beds.

Control: Most broadleaf herbicides are effective

Similar Species: mouse-ear chickweed (page 11), thymeleaf speedwell (page 34)

Chenopodiaceae (Goosefoot Family)

Chenopodium album—Common Lambsquarters

- Summer annual broadleaf
- Leaves are alternately arranged, egg-shaped, and irregularly toothed
- Young are white to gray
- Small, green flowers in dense clusters

Habitat: Disturbed sites and near agricultural fields

Control: Most broadleaf herbicides are effective

Similar Species: Other *Chenopodium* spp.

Commelinaceae (Dayflower Family)

Commelina communis—Asiatic Dayflower

Other Common Name: common dayflower

- Summer annual spiderwort
- Thick, fleshy leaves lack petioles; spreads by rooting at nodes
- Two large blue petals above one small white petal
- Dayflowers are monocots that resemble dicots

Habitat: Usually in landscape beds but may be in turf

Control: Most broadleaf herbicides are effective

Similar Species: deer-tongue grass (page 23), spreading dayflower (*Commelina diffusa*)

Convolvulaceae (Bindweed Family)

Convolvulus arvensis—Field Bindweed

Other Common Name: small bindweed

- Perennial broadleaf
- Arrowhead-shaped leaves with white to pink morningglory-like flowers
- Spreads by aboveground vines and belowground rhizomes

Habitat: Dry soils with thin turf

Control: Most broadleaf herbicides (including quinclorac) are effective when applied in fall

Similar Species: hedge bindweed (*Calystegia sepium*)

Cyperaceae (Sedge Family)

Carex blanda—Woodland Sedge

- Perennial sedge
- Tufted sedge species found in shaded lawns
- Lacks rhizomes and tubers
- Three-ranked leaves

Habitat: Shaded areas

Control: Increasing turf density and reducing shade are effective

Similar Species: *Carex* spp., yellow nutsedge (page 13)

Cyperus compressus—Annual Sedge

Other Common Name: poorland flatsedge

- Summer annual sedge
- Tufted, three-ranked leaves with smooth stems giving rise to seedheads
- Lacks rhizomes and tubers
- Seedheads are flattened, digitate clusters

Habitat: Compacted, wet areas with thin turf

Control: See *Sedge Control and Turfgrass Tolerance Ratings*, page 99

Similar Species: shining flat sedge (page 13)

Cyperus bipartitus—Shining Flat Sedge

Other Common Name: slender flatsedge

- Summer annual sedge
- Tufted, three-ranked leaves with smooth stems giving rise
- to seedheads
- Lacks rhizomes and tubers
- Colorful seedheads are primarily purplish-brown or reddish-brown

Habitat: Wet areas

Control: Similar to annual sedge; see *Sedge Control and Turfgrass Tolerance Ratings*, page 99

Similar Species: annual sedge (page 13)

Cyperus esculentus—Yellow Nutsedge

Other Common Name: yellow nutgrass

- Perennial sedge
- Three-ranked, yellow-green, shiny leaves with a distinct midrib
- Triangular stems give rise to a yellow seedhead
- Has a more sharply pointed leaf tip than purple nutsedge

Habitat: Full-sun areas; favors moist soils but also grows on well-drained sites.

Control: See *Yellow Nutsedge*, page 55

Similar Species: purple nutsedge (page 14), woodland sedge

Cyperus rotundus—Purple Nutsedge

- Perennial sedge
- Three-ranked, dark green, shiny leaves
- Triangular stems give rise to a purple seedhead

Habitat: Full-sun areas; favors moist soils but also grows on well-drained sites; less cold hardy than yellow nutsedge.

Control: See *Purple Nutsedge*, page 53

Similar Species: yellow nutsedge (page 13)

Kyllinga gracillima—False-Green Kyllinga

Other Common Names: Asiatic greenhead, pasture spikesedge

- Perennial sedge
- Tufted (three-ranked leaves) sedge species with rhizomes
- Green flowers resemble small burs

Habitat: Wet, close-mown areas

Control: See *Sedge Control and Turfgrass Tolerance Ratings*, page 99

Similar Species: green kyllinga (*Kyllinga brevifolia*), tufted kyllinga (*Kyllinga pumila*)

Equisetaceae (Horsetail Family)

Equisetum hyemale—Scouringrush

Other Common Name: rough horsetail

- Perennial horsetail
- Rhizomatous perennial with evergreen, tubelike stems
- Ornamental plant that can spread into lawns

Habitat: Generally found in wet areas but can tolerate a range of conditions

Control: One of the more difficult to control weeds; usually requires multiple glyphosate applications to fresh-cut stems; use MCPA ester in cool-season turf or metsulfuron in warm-season turf; use dichlobenil in landscape beds

Similar Species: field horsetail (*Equisetum arvense*)

Euphorbiaceae (Spurge Family)

Euphorbia maculata (Chamaesyce maculata, Euphorbia supina)—Prostrate Spurge

Other Common Names: spotted spurge, spotted sandmat

- Summer annual broadleaf
- Opposite, green leaves, often with a maroon blotch
- Hairy, pinkish stems exude milky sap when broken
- Small white flowers with pink, three-lobed seed capsules

Habitat: Turf thinned by summer stress

Control: Most broadleaf herbicides are effective

Similar Species: prostrate knotweed (page 29)

Fabaceae (Pea Family)

Lespedeza striata (Kummerowia striata)—Lespedeza

Other Common Name: Japanese clover

- Summer annual broadleaf
- Prominent veins on trifoliate leaves; each trifoliate leaf has a short petiole
- Forms mats with tough stems and taproots
- Small purple flowers

Habitat: Thin, dry, nitrogen-deficient areas in southern North Central states

Control: Fluroxypyr or triclopyr are effective

Similar Species: Other members of the pea family

Lotus corniculatus—Birdsfoot Trefoil

- Perennial broadleaf
- Oblanceolate-shaped trifoliate leaves
- Spreads by rhizomes and stolons
- Large, yellow, pealike flower with pealike fruit pods in the form of a bird's foot

Habitat: Tolerates a range of conditions, including drought; indicative of low-fertility

Control: Several broadleaf herbicides are effective, especially triclopyr

Similar Species: Other members of the pea family

Medicago lupulina—Black Medic

- Winter or summer annual broadleaf
- Trifoliate leaves; the central leaf has a longer petiole than the other leaves
- Small cluster of yellow flowers that mature into a black fruit cluster

Habitat: Tolerates a range of conditions, including drought; indicative of low-fertility

Control: Most broadleaf herbicides are effective

Similar Species: Other members of the pea family; large hop clover (page 16) is similar but its seeds and flower clusters turn light brown at maturity

Securigera varia (Coronilla varia)—Crownvetch

Other Common Names: purple crownvetch, trailing crownvetch

- Perennial broadleaf
- Dense, leafy vegetation with hairless, compound leaves
- Unlike other vetches, crownvetch lacks tendrils
- Pealike, pink and white flowers similar in shape to birdsfoot trefoil

Habitat: Commonly found along highway rights-of-way

Control: 2,4-D, clopyralid, and triclopyr are effective

Similar Species: other vetches (*Vicia* spp.)

Trifolium campestre—Large Hop Clover

Other Common Names: field clover, hop trefoil, low hop clover

- Winter annual broadleaf
- Trifoliate leaves; the central leaf has a longer petiole than the other leaves
- Small cluster of yellow flowers
- Flower petals mature turning light brown

Habitat: Thin turf, roadsides, pastures, etc.

Control: Most broadleaf herbicides are effective

Similar Species: black medic (page 15), lespedeza (page 15)

Trifolium pratense—Red Clover

- Perennial broadleaf
- Trifoliate leaves with a wedge-shaped mark; each trifoliate leaf has a short petiole
- Flowers are purple or rose-purple
- Stems are hairy, unlike white clover

Habitat: Indicative of low fertility

Control: See *White Clover*, page 55

Similar Species: Other members of the pea family. Red clover is larger than white clover (page 16).

Trifolium repens—White Clover

Other Common Name: Dutch clover

- Perennial broadleaf
- Trifoliate leaves with or without a wedge-shaped mark; each trifoliate leaf has a short petiole
- Flowers are white and turn pink as they age

Habitat: Indicative of low fertility; tolerates close mowing.

Control: See *White Clover*, page 55

Similar Species: Other members of the pea family

Geraniaceae (Geranium Family)

Erodium cicutarium—Redstem Filaree

Other Common Name: redstem stork's bill

- Winter annual or biennial broadleaf
- A prostrate, rosette-forming broadleaf with compound, lobed, hairy leaves
- Pink to purple flowers form in clusters
- Fruit has the appearance of a beak or stork's bill atop hairy stems

Habitat: Found in turf and landscapes

Control: Most broadleaf herbicides should be effective

Similar Species: Carolina geranium (page 17), musky stork's bill (*Erodium moschatum*)

Geranium carolinianum—Carolina Geranium

Other Common Name: Carolina crane's-bill

- Usually biennial or winter annual broadleaf
- Leaves are hairy, rounded, or kidney-shaped, and deeply palmately divided
- Flowers are present in late spring and summer with whitish-pink to purple flowers
- Fruit has the appearance of a crane's bill

Habitat: Often found in thin turf, shady sites

Control: Most broadleaf herbicides are effective

Similar Species: dovefoot geranium (page 17)

Geranium molle—Dovefoot Geranium

Other Common Name: dove's foot geranium

- Winter annual broadleaf
- Low and spreading with rounded or kidney-shaped leaves that are palmately divided (3–5 segments)
- Flowers are present in late spring with pink flowers
- Fruit has the appearance of a crane's bill

Habitat: Disturbed areas, roadsides, moist, shaded ground in lawns

Control: Most broadleaf herbicides are effective

Similar Species: Carolina geranium, but dovefoot geranium leaves are less deeply lobed than Carolina geranium (page 17)

Juncaceae (Rush Family)

Juncus tenuis—Path Rush

Other Common Name: slender rush

- Perennial rush
- Rush species with grasslike, slender leaves and a bunch-type growth habit
- Thin stems are round, solid, and give rise to small (about 3 mm long), egg-shaped seed capsules

Habitat: Often found in compacted soils

Control: Difficult to control with herbicides; high label rates of 2,4-D ester in cool-season turf are somewhat effective; bentazon in warm-season turf is most effective

Similar Species: Other rushes, annual sedge (page 13)

Lamiaceae (Mint Family)

Glechoma hederacea—Ground Ivy

Other Common Names: alehoof, creeping Charlie

- Perennial broadleaf
- Kidney-shaped leaves with rounded teeth on edges
- Belongs to the mint family, has square stems, and gives a mint-like odor when mown
- Spreads by stolons; has tube-shaped, purplish-blue flower petals

Habitat: Shaded, moist areas, but tolerates full sun

Control: Fluroxypyr or triclopyr are effective

Similar Species: common mallow (alternate leaves) (page 19), henbit (lacks stolons) (page 18)

Lamium amplexicaule—Henbit

- Winter annual broadleaf
- Kidney-shaped leaves, oppositely arranged, with rounded teeth on the leaf margin; upper leaves surround the stem
- Belongs to the mint family and has square stems
- Tube-shaped, pink to purple flower petals

Habitat: Disturbed, moist areas in lawns and landscape beds; more common in warm-season turf.

Control: Most broadleaf herbicides are effective; young plants are easier to control in fall.

Similar Species: purple deadnettle (triangular leaves) (page 18), ground ivy (stoloniferous) (page 17)

Lamium purpureum—Purple Deadnettle

- Winter annual broadleaf
- Trianglar leaves are oppositely arranged; upper leaves crowd the stem and may be purple
- Belongs to the mint family and has square stems
- Tube-shaped, light purple flower petals

Habitat: Disturbed, moist areas in lawns and landscape beds; more common in warm-season turf.

Control: Most broadleaf herbicides are effective; young plants are easier to control in fall.

Similar Species: henbit (page 18)

Prunella vulgaris—Healall

Other Common Name: self heal

- Perennial broadleaf
- Oppositely arranged, egg-shaped leaves on petioles
- Belongs to the mint family and has square stems
- Tube-shaped, light purple flower petals surround a dense spike at the end of an ascending stem

Habitat: Disturbed, moist areas in lawns

Control: Quinclorac, fluroxypyr, and triclopyr are effective

Similar Species: Not often confused with other weeds, although the flowers are similar to other Lamiaceae weeds

Liliaceae (Lily Family)

Allium canadense—Wild Onion

Other Common Name: meadow garlic

- Perennial lily
- A bulbous, grass-like plant that emerges in late winter and early spring
- Leaves are linear, smooth, flat in cross section, and not hollow
- Flowers in late spring

Habitat: Tolerates a range of soil conditions

Control: See *Wild Garlic,* page 55

Similar Species: wild garlic (page 19), star-of-Bethlehem (page 19)

Allium vineale—Wild Garlic

Other Common Name: field garlic

- Perennial lily
- A bulbous, grasslike plant that emerges in late winter and early spring
- Produces flowers or aerial bulblets in late spring

Habitat: Tolerates a range of soil conditions

Control: See *Wild Garlic*, page 55

Similar Species: wild onion (page 18), star-of-Bethlehem (page 19)

Ornithogalum umbellatum—Star-of-Bethlehem

- Perennial lily
- A bulbous, grasslike plant found in early spring
- Leaves are linear, smooth, flat in cross section, and have a white midrib
- Showy, 6-petaled white flowers have a distinctive green stripe underneath and appear in spring

Habitat: Shaded, moist areas

Control: Sulfentrazone and carfentrazone are effective

Similar Species: crowpoison (*Nothoscordum bivalve*), spring beauty (page 31), wild garlic (page 19), wild onion (page 18)

Malvaceae (Mallow Family)

Malva neglecta—Common Mallow

- Winter or summer annual broadleaf
- Alternately arranged, circular to kidney-shaped leaves with shallow lobes and toothed margins
- White or whitish-lavender, five-petaled flowers

Habitat: Low-maintenance turf and landscapes

Control: Most broadleaf herbicides are effective

Similar Species: ground ivy (opposite leaves) (page 17)

Sida spinosa—Prickly Sida

Other Common Names: prickly fanpetals, teaweed

- Summer annual broadleaf
- Oblong leaves borne on woody stems have serrated margins
- Small spines are noticeable at stem nodes
- Pale yellow, 5-petaled flowers

Habitat: Found in fields, pastures, and low-input turf

Control: Use herbicides with fluroxypyr or triclopyr; as the weeds get older, control with herbicides will decline

Similar Species: white mulberry (*Morus alba*)

Molluginaceae (Carpetweed Family)

Mollugo verticillata—Carpetweed

- Summer annual broadleaf
- Prostrate mats of stems along the soil surface with whorled leaves (3–8) at each stem node
- Small white flowers in late summer

Habitat: Thin turf, spring- and summer-seeded areas

Control: Most broadleaf herbicides are effective

Similar Species: smooth bedstraw (page 33), catchweed bedstraw (page 33)

Oxalidaceae (Woodsorrel Family)

Oxalis stricta—Yellow Woodsorrel

Other Common Name: common yellow oxalis

- Perennial or summer annual broadleaf
- A cloverlike plant with trifoliate, heart-shaped leaflets
- Flowers are yellow with five petals
- Primarily spreads by seeds; seeds are housed in a cylindrical capsule that explodes to ejects seeds when mature

Habitat: Tolerates a range of soil conditions

Control: Fluroxypyr and triclopyr are effective

Similar Species: black medic (page 15), birdsfoot trefoil (page 15), white clover (page 16)

Plantaginaceae (Plantain Family)

Plantago lanceolata—Buckhorn Plantain

Other Common Name: narrowleaf plantain

- Perennial broadleaf
- Rosette growth habit with lanceolate to elliptic-shaped leaves and prominent parallel venation
- Leaf blades are smooth but may have silky hairs at base
- Flowers and seedheads borne at the top of rigid stalks

Habitat: Dry, low fertility soils

Control: Control with 2,4-D is variable; clopyralid is effective; fall applications work best.

Similar Species: broadleaf plantain (page 20), bracted plantain (*Plantago aristata*)

Plantago major—Broadleaf Plantain

Other Common Names: comman plantain, ribwort

- Perennial broadleaf
- Rosette growth habit with elliptic to oval-shaped leaves
- Leaf blades usually smooth with light green or reddish-green petioles; leaf surfaces appear waxy
- Seeds borne almost entirely along the flower stalk

Habitat: More often found on rich, moist soils

Control: 2,4-D provides best control

Similar Species: blackseed plantain (page 21), bracted plantain, buckhorn plantain (page 20)

Plantago rugelii—Blackseed Plantain

Other Common Name: Rugel's plantain

- Perennial broadleaf
- Rosette growth habit with elliptic to oval-shaped, pleated leaves and prominent parallel venation
- Larger leaves than broadleaf plantain have red or purplish petioles; leaf surfaces appear waxy.
- Seeds borne nearly entirely along the flower stalk

Habitat: Tolerates a range of soil conditions, but often found on rich, moist soils

Control: Use herbicides with 2,4-D for best control

Similar Species: bracted plantain, broadleaf plantain (page 20), buckhorn plantain (page 20)

Poaceae (Grass Family)

Agrostis gigantea—Redtop

- Perennial grass
- Rhizomatous species with large membranous ligule
- More erect with wider leaves than creeping bentgrass
- Pyramid-shaped panicle that is reddish

Habitat: Disturbed sites such as fields, pastures, roadsides, and lawns planted with poor seed mixtures

Control: Difficult to control; glyphosate, fluazifop, or possibly mesotrione are most effective.

Similar Species: creeping bentgrass (page 20), quackgrass (page 24)

Agrostis stolonifera—Creeping Bentgrass

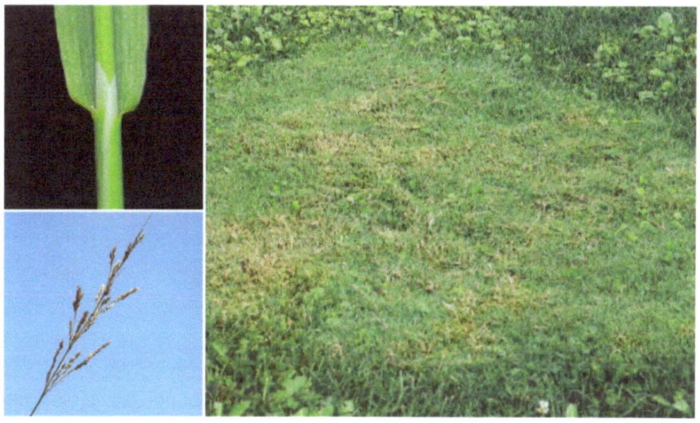

- Perennial grass
- Stoloniferous, creeping *Agrostis* spp. with a large membranous ligule
- Prostrate growth habit; forms patches in lawns
- Subject to summertime scalping from rotary mowers

Habitat: Found in mown and unmown areas, particularly on golf courses; prefers rich, moist soils.

Control: See *Creeping Bentgrass*, page 48

Similar Species: nimblewill (short ligule) (page 25), redtop (rhizomes) (page 21)

Bromus inermis—Smooth Brome

Other Common Name: awnless brome

- Perennial grass
- Rhizomatous grass often found on roadsides
- A short ligule, no auricle, and smooth leaf blade and sheath; the leaf may have distinct W-shaped mark
- Leaf sheath is split halfway down the stem

Habitat: Abundant in pastures and roadsides, particularly in the central United States

Control: Difficult to control; multiple applications of glyphosate (for control) or mesotrione (for suppression).

Similar Species: quackgrass (page 24)

Cenchrus longispinus—Sandbur

Other Common Names: longspine sandbur, mat sandbur

- Summer annual grass
- Prostrate growth habit with some rooting at nodes
- Smooth leaf blade; leaf sheath has hairy margins
- The seedhead contains bur-like fruit with several long spines

Habitat: Usually found on sandy soils in the Midwest

Control: See *Sandbur*, page 54

Similar Species: yellow foxtail (page 28) or crabgrass before flowering; field sandbur (*C. incertus*) is a similar species that appears more in the South

Chloris verticillata—Windmillgrass

Other Common Name: tumble windmillgrass

- Perennial grass
- Prostrate, stoloniferous grass with compressed stems
- Leaves are gray to bluish-green
- Has a distinct, large, verticillate panicle seedhead; the seedhead breaks away at maturity and rolls in the wind as a tumbleweed

Habitat: Dry, compacted soils next to roadsides, parking lots, sidewalks

Control: Fenoxaprop, mesotrione, and topramezone are effective

Similar Species: bermudagrass (page 22), large crabgrass (page 23)

Cynodon dactylon—Common Bermudagrass

Other Common Names: couchgrass, devil's grass, green couch

- Perennial grass
- Spreads by rhizomes and stolons to form dense mats
- Leaves are smooth or hairy; ligule is hairy; seedhead is digitate with 3–6 fingerlike spikes
- Bermudagrass holds a heavy dew in the morning. Stolon tips have leaves unlike leafless stolon tips of zoysiagrass

Habitat: Tolerates a wide range of soil conditions including drought; intolerant of shade.

Control: See *Bermudagrass*, page 43

Similar Species: nimblewill (page 25), zoysiagrass (page 29)

Dactylis glomerata—Orchardgrass

Other Common Name: cocksfoot

- Perennial grass
- Bunch-type growth habit with upright leaves
- Leaves are bluish-green with a long ligule and boat-shaped tip; sheaths are compressed
- The seedhead is a branched panicle

Habitat: A forage species often found in lawns. It is also found in meadows and roadsides.

Control: See *Orchardgrass*, page 52

Similar Species: annual bluegrass (page 27), tall fescue (page 28)

Dichanthelium clandestinum (Panicum clandestinum)—Deertonguegrass

Other Common Names: deertongue, hidden panic grass

- Perennial grass
- Dense colonies form from short rhizomes
- Lance-shaped leaves are 4–7 inches long and 1 inch wide; leaf sheaths are hairy
- Seedhead is panicle with a pyramidal shape and spikelets

Habitat: Often found in moist or wet sites

Control: Glyphosate, fluazifop, topramezone, and imazapic are most effective

Similar Species: Asiatic dayflower (page 12), Japanese stiltgrass (page 25)

Digitaria ischaemum—Smooth Crabgrass

Other Common Name: watergrass

- Summer annual grass
- Prostrate growth habit as plants tiller (after producing 4 to 5 leaves) in summer
- Leaf blades and sheaths are smooth although a few fine hairs may be present; ligule is membranous.
- Seedhead is digitate with 3–5 fingerlike spikes

Habitat: Tolerates a range of conditions and is often found in turf

Control: See *Crabgrass*, page 45

Similar Species: large crabgrass (page 23), yellow foxtail (page 28), barnyardgrass (page 23)

Digitaria sanguinalis—Large Crabgrass

Other Common Name: hairy crabgrass

- Summer annual grass
- Prostrate growth habit as plants tiller (after producing 4 to 5 leaves) in summer
- Leaf blades and sheaths are hairy; ligule is membranous
- Seedhead is digitate with 3–5 fingerlike spikes

Habitat: Tolerates a range of conditions and is found almost anywhere

Control: See *Crabgrass*, page 45

Similar Species: smooth crabgrass (page 23), yellow foxtail (page 28)

Echinochloa crus-galli—Barnyardgrass

- Summer annual grass
- Prostrate growth habit in turf, but erect when unmown
- Leaf blades and sheaths are smooth; ligule is absent; midrib is distinct
- Seedhead is branched and seeds are with or without awns

Habitat: Usually found in moist soils. Often found in turf near agricultural areas.

Control: Similar to Crabgrass. See *Crabgrass*, page 45

Similar Species: smooth crabgrass (page 23), fall panicum (page 26), giant foxtail (page 28), johnsongrass (page 29)

Eleusine indica—Goosegrass

Other Common Name: silver crabgrass

- Summer annual grass
- Prostrate growth habit as plants tiller (after producing less than 6 leaves) in summer
- Leaves are folded in vernation; sheaths are compressed and light green at the base
- Seedhead is digitate with 3–8 fingerlike spikes; seeds are flattened and arranged in two rows like a zipper

Habitat: Tolerates close mowing and often found in compacted soils

Control: See *Goosegrass*, page 50

Similar Species: smooth crabgrass (page 23), large crabgrass (page 23)

Elymus repens (Agropyron repens, Elytrigia repens)—Quackgrass

- Perennial grass
- A rhizomatous grass with wide leaves and a long, clasping auricle
- Leaves are rolled in vernation; sheaths are smooth or may have short hairs
- Seedhead is a long spike with seed arranged in two rows

Habitat: Found in Midwest turf and landscapes

Control: See *Quackgrass*, page 53

Similar Species: annual ryegrass (page 25), redtop (page 21), smooth brome (page 21), tall fescue (page 28)

Eragrostis cilianesis—Stinkgrass

- Summer annual grass
- Bunch-type growth habit; Leaves rolled in bud
- Leaves sparsely hairy with tuft of hairs at the collar margin
- Seedhead is an open panicle

Habitat: Disturbed places in both dry and moist soils

Control: Seldom a problem in dense turf

Similar Species: tufted lovegrass (page 24)

Eragrostis pectinacea—Tufted Lovegrass

- Summer annual grass
- Bunch-type growth habit; leaves narrow and rolled in bud
- Leaves smooth with tuft of hairs at the collar margin
- Seedhead is an open panicle; spikelets smaller than stinkgrass

Habitat: Disturbed places in both dry and moist soils

Control: Seldom a problem in dense turf

Similar Species: stinkgrass (page 24)

Lolium multiflorum—Annual Ryegrass

Other Common Name: Italian ryegrass

- Winter annual grass
- An erect, bunch-type grass with a light green foliage
- Long, clasping auricles at the base of wide leaf blades; stem bases are often purplish-red
- Seedhead is a long spike with an alternating seed arrangement; seeds will have awns

Habitat: Lawns planted with poor seed mixtures

Control: Plant high-quality seed mixtures without annual ryegrass, Italian ryegrass, or intermediate ryegrass; no selective control options are available in cool-season turf.

Similar Species: quackgrass (page 24)

Microstegium vimineum—Japanese Stiltgrass

- Summer annual grass
- An invasive species from Asia, now spreading into the Northeast and Midwest from Appalachia
- Leaves are short (1 to 4 inches), wide (0.5 inch), and have a slivery, white, off-center midrib
- Seedheads are similar to crabgrass

Habitat: Tolerates a range of conditions including shade; found in lawns and fields, and along trails

Control: Preemergence grass herbicides are effective; fenoxaprop and MSMA are effective postemergence

Similar Species: deertonguegrass (page 23, other native grass species

Muhlenbergia schreberi—Nimblewill

- Perennial grass
- A stoloniferous grass that forms patches in lawns and is easily seen after a frost when it turns straw-colored
- Short, dagger-shaped leaves have a membranous ligule and sparse hairs on the collar
- The spiked seedhead has seeds with individual hairs (awns) on the tops of the seed

Habitat: Tolerates a range of conditions including shade and is often found in turf

Control: See *Nimblewill*, page 52

Similar Species: creeping bentgrass (page 21), bermudagrass (page 22)

Panicum capillare—Witchgrass

Other Common Name: witch panicgrass

- Summer annual grass
- An erect or sprawling, bunch-type grass with wide leaf blades
- The ligule is a fringe of hairs, but leaf blades and sheaths are very hairy
- The large seedhead is a branched panicle

Habitat: Usually found in cultivated and disturbed areas

Control: Similar to Crabgrass; see *Crabgrass*, page 45

Similar Species: fall panicum (page 26)

Panicum dichotomiflorum—Fall Panicum

Other Common Name: fall panicgrass

- Summer annual grass
- An erect or sprawling, bunch-type grass with wide leaf blades
- The ligule is a fringe of hairs but leaf blades and sheaths are smooth
- The large seedhead is a branched panicle

Habitat: Usually found in cultivated and disturbed areas

Control: Similar to Crabgrass; see *Crabgrass*, page 45

Similar Species: witchgrass (page 25)

Paspalum dilatatum—Dallisgrass

- Perennial grass
- Wide-bladed (6–15 mm wide) grass easily visible in mid- and late summer
- Has short rhizomes near the soil surface, but spreads primarily by seed
- Panicles with racemosely arranged branches (3–8); seeds are hairy and in four rows.

Habitat: Found in pastures, roadsides, and turf (tolerates close mowing); less winter hardy than field paspalum.

Control: See *Dallisgrass*, page 49

Similar Species: field paspalum (spikelets in two rows) (page 26), large crabgrass (page 23), purpletop (page 29), thin paspalum (*Paspalum setaceum*)

Paspalum laeve—Field Paspalum

Other Common Name: smooth lensgrass

- Perennial grass
- Wide-bladed (6–15 mm wide) grass seen in mid- and late summer
- Has short rhizomes, but spreads primarily by seed
- Panicles with racemosely arranged branches (2–7); seeds smooth and in two rows

Habitat: Found in pastures, roadsides, and turf (tolerates close mowing)

Control: See *Dallisgrass*, page 49

Similar Species: dallisgrass (page 26), large crabgrass (page 23), purpletop (page 29), thin paspalum (*Paspalum setaceum*)

Pennisetum alopecuroides—Fountain Grass

Other Common Name: Chinese fountain grass

- Perennial grass
- Bunch-type growth habit; hairs visible on the leaf collar
- Seedhead is a spikelike raceme, which is cylindrical and resembles foxtails
- This escaped ornamental grass can be identified by its shredded leaf blades in summer, leaving a whitish, wispy-looking clump after mowing

Habitat: Found in lawns near landscape beds that contain fountain grass

Control: See *Fountain Grass*, page 50

Similar Species: foxtails, annual ryegrass (page 25)

Phleum pretense—Timothy

Other Common Name: herd's grass

- Perennial grass
- Rhizomatous grass with an upright habit; stems are swollen at the base
- Leaves have a long membranous ligule and pointed tip
- The seedhead is a spikelike panicle

Habitat: Timothy is a forage species often found in lawns

Control: Sethoxydim is the most effective selective herbicide (depending on turf species and site use)

Similar Species: foxtails, quackgrass (page 24), redtop (page 21)

Poa bulbosa—Bulbous Bluegrass

Other Common Name: bulbous meadowgrass

- Perennial grass
- Bunch-type growth habit with swollen stem bases (bulbs); this grass is found in early spring
- Leaves medium green, folded in bud, and have a boat-shaped tip and a membranous ligule
- Seedhead is a pyramid-shaped panicle with aerial bulblets

Habitat: Prefers moist, cool, shaded conditions

Control: Increasing turf density and reducing shade are effective

Similar Species: other bluegrasses

Poa annua—Annual Bluegrass

Other Common Name: annual meadowgrass

- Winter annual or weak perennial grass
- Bunch-type growth habit with occasional stolons
- Leaves are yellowish-green, folded in bud, and have a boat-shaped tip and a membranous ligule
- Seedhead is a pyramid-shaped panicle

Habitat: Prefers moist, cool, shaded conditions; grows well at short cutting heights and in compacted areas.

Control: See *Annual Bluegrass*, page 42

Similar Species: other bluegrasses

Poa trivialis—Rough Bluegrass

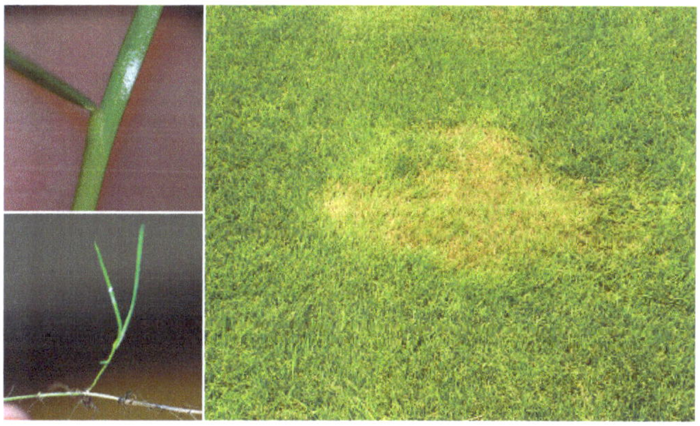

Other Common Names: roughstalk bluegrass, roughstalk meadowgrass

- Perennial grass
- Spreads by stolons to form patches; heat and drought intolerant
- Leaves are metallic green, folded in bud, and have a boat-shaped tip; ligule is typically short; senescing leaves appear like onion skin; sheath is split and may be purple
- Quickly enters dormancy and appears dead in hot summers

Habitat: Prefers moist, cool, shaded conditions

Control: See *Rough Bluegrass*, page 54

Similar Species: other bluegrasses

Schedonorus arundinaceus—Tall Fescue

- Perennial grass
- Bunch-type growth habit; leaves are wide and have prominent veins, short auricles, a short ligule, and a broad collar
- Clumps most visible in early spring and summer drought
- Former scientific name: *Festuca arundinacea*

Habitat: Tolerates a range of conditions, including drought

Control: See *Tall Fescue*, page 54

Similar Species: annual ryegrass (page 25), quackgrass (page 24)

Setaria faberi—Giant Foxtail

- Summer annual grass
- Bunch-type growth habit; larger than yellow or green foxtails
- Leaves are rolled in bud; leaf blades have several short hairs
- Bottlebrush or fox tail-like seedheads; seedhead is nodding or droopy

Habitat: Found in pastures, roadsides, and unmown turf

Control: Similar to crabgrass. See *Crabgrass*, page 23

Similar Species: green foxtail (page 28), fountain grass (page 26)

Setaria glauca—Yellow Foxtail

- Summer annual grass
- Bunch-type growth habit
- Leaves are rolled in bud; leaf blades have several long hairs at the base of the upper leaf surface
- Bottlebrush or fox taillike seedheads

Habitat: Found in pastures, roadsides, and turf (tolerates close mowing)

Control: Similar to crabgrass; see *Crabgrass*, page 45

Similar Species: green foxtail (page 28), fountain grass (page 26)

Setaria viridis—Green Foxtail

- Summer annual grass
- Bunch-type growth habit
- Leaves are rolled in bud; leaf blades lack hairs
- Bottlebrush or fox taillike seedheads

Habitat: Found in pastures, roadsides, and turf (tolerates mowing)

Control: Similar to crabgrass; see *Crabgrass*, page 45

Similar Species: yellow foxtail (page 28), fountain grass (page 26)

Sorghum halepense—Johnsongrass

- Perennial grass
- Wide-bladed grass with a distinct midrib; spreads by rhizomes
- Ligule is a fringe of hairs; leaves are mostly smooth
- The seedhead is a large, purplish panicle; some (but not all) seeds have a twisted awn

Habitat: Roadsides and low-maintenance turf areas

Control: This is a noxious weed in most states; glyphosate, MSMA, and sulfosulfuron are most effective (depending on turf species and site use)

Similar Species: purpletop (page 29)

Tridens flavus var. *flavus*—Purpletop

Other Common Name: purpletop tridens

- Perennial grass
- Wide-bladed grass, slightly smaller than Johnsongrass, with a bunch-type growth habit
- Ligule is a fringe of hairs and a tuft of hair is visible on each side of the collar; the seedhead is a large, purple panicle; seeds lack an awn

Habitat: Roadsides and low-maintenance turf areas

Control: This native species is often considered desirable; mechanical removal is most effective.

Similar Species: johnsongrass (page 29), thin paspalum (*Paspalum setaceum*)

Zoysia japonica—Zoysiagrass

Other Common Name: Japanese lawngrass

- Perennial grass
- Spreads by rhizomes and stolons, and forms dense mats
- Leaves are sparsely hairy; ligule is hairy; seedhead is spikelike
- Turns golden brown in winter

Habitat: Tolerates a wide range of soil conditions, including drought and moderate shade

Control: See *Zoysiagrass*, page 56

Similar Species: bermudagrass (page 22), nimblewill (page 25)

Polygonaceae (Buckwheat Family)

Polygonum aviculare—Prostrate Knotweed

- Summer annual broadleaf
- Prostrate growth habit with branched stems
- Leaves are alternately arranged; similar to other Polygonaceae family weeds, an ocrea (membranous sheath) surrounds the stem at the base of the leaf
- Small white flowers are barely visible

Habitat: Found in compacted soils next to drives and sidewalks

Control: See *Prostrate Knotweed*, page 53

Similar Species: prostrate spurge (page 14)

Polygonum pensylvanicum—Pennsylvania Smartweed

Other Common Name: swamp persicary

- Summer annual broadleaf
- Erect, highly branched plant with alternately arranged, lanceolate leaves
- Bright pink to white flowers in dense, spikelike clusters that bloom in summer
- A membranous sheath (ocrea) surrounds the stem

Habitat: Often found in moist soils

Control: Most broadleaf herbicides are effective

Similar Species: ladysthumb (fringed ocrea) (page 30)

Polygonum persicaria—Ladysthumb

Other Common Name: spotted ladysthumb

- Summer annual broadleaf
- Has sprawling to erect growth habit; a highly-branched plant with alternately arranged lanceolate leaves; has dark pigment mark that is usually missing on Pennsylvania smartweed
- Bright pink to white flowers in dense, spikelike clusters that bloom in summer
- A fringed ocrea surrounds the stem

Habitat: Often found in moist soils

Control: Most broadleaf herbicides are effective

Similar Species: Pennsylvania smartweed (page 30)

Rumex acetosella—Red Sorrel

Other Common Name: common sheep sorrel

- Perennial broadleaf
- Spreads by rhizomes, low-growing, and has unique, arrowhead-shaped leaves
- Red to reddish-brown female flowers are produced on a branched seedhead

Habitat: Tolerates a range of conditions; often more predominant in acid soils.

Control: Most broadleaf herbicides are effective

Similar Species: field bindweed (page 12)

Rumex crispus—Curly Dock

Other Common Name: yellow dock

- Perennial broadleaf
- Rosette growth habit with wavy-margined leaves
- Flowers rarely seen in turf but turn from green to reddish-brown at maturity

Habitat: Thin turf, roadsides, pastures, etc.

Control: Most broadleaf herbicides are effective

Similar Species: broadleaf dock (*Rumex obtusifolius*)

Portulacaceae (Purslane Family)

Claytonia virginica—Spring Beauty

Other Common Name: Virginia springbeauty

- Perennial broadleaf
- An ephemeral plant often found in shady areas
- Leaves are green and somewhat grasslike
- Showy, 5-petaled, white flowers with pink or light purple stripes appear in early spring

Habitat: Shaded and partially shaded lawns

Control: Increase turf density and reduce shade

Similar Species: star-of-Bethlehem (page 19)

Portulaca oleracea—Common Purslane

Other Common Name: little hogweed

- Summer annual broadleaf
- Prostrate, mat-forming growth habit
- Leaves and stems are thick and succulent; stems are often purplish-red
- Flowers are small and yellow (5-petals); seeds are black

Habitat: Thin turf, spring- and summer-seeded areas

Control: Dicamba, fluroxypyr, metsulfuron, and triclopyr are most effective

Similar Species: prostrate spurge (page 14), prostrate knotweed (page 29)

Primulaceae (Primrose Family)

Anagallis arvensis—Scarlet Pimpernel

- Winter or summer annual broadleaf
- Low and spreading plant with triangular to heart-shaped leaves
- Stems are branched and square in cross section
- Flowers (5-petals) are salmon to orange

Habitat: Thin turf, roadsides, pastures, etc.

Control: Most broadleaf herbicides are effective

Similar Species: common chickweed (round stems) (page 12)

Lysimachia nummularia—Moneywort

Other Common Names: creeping Jenny, creeping loosestrife

- Perennial broadleaf
- A nonclimbing vine occasionally sold as an ornamental
- Oppositely arranged, shiny, round leaves
- Bright yellow flowers in late spring and late summer

Habitat: Prefers moist, shady sites and tolerates close mowing

Control: Most broadleaf herbicides should be effective

Similar Species: thymeleaf speedwell (page 34), ground ivy (page 17)

Ranunculaceae (Buttercup Family)

Ranunculus ficaria—Lesser Celandine

Other Common Name: fig buttercup

- Perennial broadleaf
- Low-growing herbaceous plant with dark-green, shiny, heart-shaped to stalked-kidney-shaped leaves
- Produces small bulbils at the bases of the stem
- Bright yellow flowers with 8–12 petals

Habitat: An invasive, shade-loving ornamental that prefers wooded, wetland areas and spreads into lawns

Control: Herbicide effectiveness largely unknown

Similar Species: marsh marigold (*Caltha palustris*)

Rosaceae (Rose Family)

Duchesnea indica—Indian Mock-Strawberry

Other Common Name: mock strawberry

- Perennial broadleaf
- Low-growing plant with dark-green, trifoliate leaves with crenate (rounded) toothed margins
- Spreads by creeping stolons
- Yellow flowers (5-petals) give rise to small red fruit

Habitat: Shaded areas in lawns, pastures, and meadows on the edge of wooded areas

Control: Most broadleaf herbicides are effective

Similar Species: wild strawberry (*Fragaria virginiana*), oldfield cinquefoil (*Potentilla simplex*)

Potentilla recta—Sulfur Cinquefoil

- Perennial broadleaf
- Erect growing broadleaf with palmately compound, hairy leaves (5–9 leaflets)
- Lacks rhizomes and stolons
- Yellow flowers borne on hairy stems

Habitat: Thin turf, roadsides, pastures, etc.

Control: Most broadleaf herbicides are effective

Similar Species: other cinquefoils (*Potentilla* spp.)

Rubiaceae (Madder Family)

Diodia virginiana—Virginia Buttonweed

- Perennial broadleaf
- Prostrate, spreading weed with dark green, elliptical, alternately arranged leaves
- Flowers (4-petals) are star-shaped
- Oval-shaped, hairy fruit capsules

Habitat: Full-sun, moist areas

Control: Difficult to control; multiple applications of herbicides with fluroxypyr or clopyralid in cool-season turf are effective; multiple applications of herbicides with metsulfuron, thiencarbazone, or trifloxysulfuron in warm-season turf are effective

Similar Species: poorjoe (*Diodia teres*)

Galium mollugo—Smooth Bedstraw

Other Common Name: false baby's breath

- Perennial broadleaf
- Mat-forming plant with smooth stems, stolons, and rhizomes
- Whorled leaves on square stems; smaller than catchweed bedstraw
- Flowers (4-petals) are star-shaped

Habitat: Thin turf, roadsides, pastures, etc.

Control: Most broadleaf herbicides are effective

Similar Species: catchweed bedstraw (page 33), carpetweed (page 20)

Gallium aparine—Catchweed Bedstraw

Other Common Name: cleavers, stickywilly

- Winter annual broadleaf
- Mat-forming plant; stems, leaves, and fruit have prickly hairs
- Whorled leaves on square stems
- Flowers (4-petals) are star-shaped

Habitat: Thin turf, fall seedings

Control: Most broadleaf herbicides are effective

Similar Species: smooth bedstraw (page 33), carpetweed (page 20)

Scrophulariaceae (Figwort Family)

Verbascum thapsus—Common Mullein

- Biennial broadleaf
- Has large rosette the first year and bolts with a stem in second year
- Leaves are densely hairy and soft
- Yellow flower petals on stalk

Habitat: Thin turf, roadsides, pastures, etc.

Control: Most broadleaf herbicides are effective

Similar Species: moth mullein (*Verbascum blattaria*)

Veronica arvensis—Corn Speedwell

- Winter annual broadleaf
- Lower stems are prostrate and leaves oppositely arranged; leaves on upright flower stalks are alternately arranged; leaves and stems are hairy
- Flowers are pale blue to white

Habitat: Thin turf, fall seedings

Control: Most broadleaf herbicides are effective

Similar Species: creeping speedwell (*Veronica filiformis*), purslane speedwell (*Veronica peregrina*)

Veronica serpyllifolia—Thymeleaf Speedwell

- Perennial broadleaf
- Lower leaves are oppositely arranged and smooth; upper leaves are alternately arranged
- Flowers are white with dark blue veins

Habitat: Thin turf, fall seedings

Control: Most broadleaf herbicides are effective

Similar Species: common chickweed (page 12), scarlet pimpernel (page 31), slender speedwell (*Veronica filiformis*), moneywort (page 31)

Solanaceae (Nightshade Family)

Solanum carolinense—Horsenettle

- Perennial broadleaf
- Spreads by rhizomes; stems and leaves have spinelike prickles
- Flowers resemble those of potato (white with yellow anthers)
- Yellow fruit contains seeds

Habitat: Thin turf, roadsides, pastures, etc.

Control: Most broadleaf herbicides are effective

Similar Species: smooth groundcherry (*Physalis subglabrata*)

Verbenaceae (Verbena Family)

Verbena bracteata—Prostrate Vervain

Other Common Name: bigbract verbena

- Summer annual broadleaf
- Prostrate growth with many branched, hairy stems; lower leaves are divided and hairy; upper leaves are smaller, hairy, and lack lobes
- Small light blue or light purple flowers bloom near the apex of dense spikes

Habitat: Disturbed sites such as fields, roadsides, and lawns; often found near gravel drives and walks

Control: Most broadleaf herbicides are effective

Similar Species: Not often confused with other weeds

Violaceae (Violet Family)

Viola sororia—Wild Violet

Other Common Name: common blue violet

- Perennial broadleaf
- Low-growing plant with heart-shaped, shiny leaves
- Plants spread by rhizomes and seed
- Typical violet-like flowers (deep purple or bluish-purple)

Habitat: Thin, shaded turf, roadsides, pastures, etc.

Control: Triclopyr is effective

Similar Species: confederate violet (page 35), English violet (*Viola odorata*), field violet (*Viola arvensis*) (yellow with purple flowers)

Viola sororia priceana—Confederate Violet

- Perennial broadleaf
- Low-growing plant with heart-shaped, shiny leaves
- Plants spread by rhizomes and seed
- Typical violet-like flowers (white with purple veins)

Habitat: Thin, shaded turf, roadsides, pastures, etc.

Control: Triclopyr is effective

Similar Species: wild violet (page 35), English violet (*Viola odorata*) (white flowers), field violet (*Viola arvensis*) (yellow with purple flowers)

Zygophyllaceae (Caltrop Family)

Tribulus terrestris—Puncturevine

Other Common Name: goathead

- Summer annual broadleaf
- Prostrate, mat-forming growth habit
- Produces many burs with sharp spines
- Small yellow flowers bear bur fruit

Habitat: Disturbed sites such as fields, pastures, roadsides, and lawns; found in sandy and compacted soils

Control: Most broadleaf herbicides are effective

Similar Species: Not often confused with other weeds

Herbicide Use

General use pesticides can be purchased and applied by anyone, including homeowners. Other products are restricted use pesticides (RUP), because there are risks involved in handling them or they have the potential to harm the environment—such products bear the words "Restricted use" on their labels.

Only licensed applicators can apply RUPs. Furthermore, most states require certain applicators to have licenses whether they are applying RUPs or general use pesticide. Commonly, licenses are required for the following:

- Any for-hire pesticide or fertilizer applicator (residential or commercial).
- Any applicator who applies pesticides to golf courses and school properties.

Check with your state pesticide regulator for exact rules, certification, and licensing information.

Use herbicides and other pesticides safely to protect against injuring or harming the applicators, turf, customers, and the environment. Always follow label directions when using herbicides, and obey all federal, state, and local pesticide laws and regulations. Labels provide specific safety suggestions and requirements for handling products.

The following are general guidelines to reduce the risks from herbicides:

- Apply a product only to the turfgrass species listed on the label.
- Clean spray tanks thoroughly when changing from one herbicide to another. Many herbicides contain instructions on how to properly clean and rinse the sprayer following an application.
- Calibrate sprayers correctly and often (see *Sprayer and Spreader Calibration*, page 41).
- Use the recommended herbicide application rates provided on the label. The label may also specify a specific rate for specific weed species. Rates listed on the manufacturer label are based on research at multiple locations across multiple years. Applying too much herbicide is costly and could result in turf damage. Applying too little herbicide can result in poor weed control and unsatisfied customers.
- Apply herbicides as specified on the label (timing, site, interval between applications, interval before and after seeding, and so on).
- Wear the appropriate personal protective equipment (PPE) specified on the label.
- Use caution when spraying around ornamental plants and sensitive crops to avoid injury. Follow wind restrictions on the label.
- Apply herbicides when temperatures are in the range provided on the labels (see *Frequently Asked Questions and Answers About Weed Control with Herbicides*, page 57).
- Do not apply herbicides when children or students are in the application area. This is known as the School Rule. More information about the School Rule is available on the Office of Inidian State Chemist OISC website, *www.oisc.purdue.edu*. Many states have similar laws—check with your state pesticide regulator. See page 124 for a listing of regulators.
- Check the label for instructions and options on how to remove pesticide residues from containers prior to their disposal.
- Store pesticides in their original containers and keep unused pesticides in a safe, secure location. Keep storage areas on trucks or within buildings locked, and keep pesticide containers away from children.

Herbicide Nomenclature

Each herbicide can be classified by three names: trade name, common name, and chemical name.

The **trade name** is a unique name assigned by the chemical company—this is often the name turf professionals and others may use.

The **common name** is a general name that is given to a specific chemical. Researchers often use the **active ingredient** as the common name. It is important to know the ingredients in a product and not just its trade name because it is useful for comparing products—especially post-patent (generic) products. Often, there are many different products sold under different trade names that contain the same active ingredient or common name.

It is also important to know common names/active ingredients because some herbicides with different ingredients have similar trade names—for example, the product with the trade name Momentum Q contains the active ingredients 2,4-D, quinclorac, and dicamba. Momentum FX2 contains 2,4-D, triclopyr, and fluroxypyr.

The **chemical name** describes the chemistry of a particular herbicide product (See *Common and Trade Names and Mechanism of Action of Registered Herbicides*, page 107, for a list of common and trade names for turf herbicides and plant growth regulators.)

Here is an example of the different names a single product can have:

- Trade Name: Dimension
- Common Name: dithiopyr
- Chemical Name: S,S-dimethyl 2-(difluoromethyl)-4-(2-methylpropyl)-6-(trifluoromethyl)-3, 5-pyridinedicarbothioate

Herbicide Classification

Herbicides can be classified as selective or nonselective and as preemergence or postemergence.

Selective and Nonselective Herbicides

Selective herbicides are products that control or suppress some plant species without harming the growth of desirable plants. Differential absorption, translocation, metabolism, and morphological and physiological differences are all causes for selectivity between turfgrasses and weeds. The majority of turf herbicides are selective. Most broadleaf herbicides (2,4-D, MCPP, dicamba, triclopyr, and more) are examples of selective herbicides because they control broadleaf weeds but don't injure turfgrasses.

Be aware that even selective herbicides can sometimes injure turf, especially when the turf is stressed from heat, drought, low mowing, or other factors. Keep in mind that selectivity is relative to many factors, including herbicide rate, environmental conditions, application timing, and the species and cultivar being treated. In other words, applying too much of a selective herbicide, or applying it at the wrong time or place, can injure turf.

For example, ethofumesate (Prograss) is a selective herbicide labeled for use on creeping bentgrass fairways. Although this herbicide is labeled as safe (good to excellent tolerance) to use in certain creeping bentgrass cultivars, the label also cautions that it may seriously injure other creeping bentgrass cultivars. In another example, some herbicides are selective to specific weed species and are used to remove one weed species within another.

Nonselective herbicides (also called broad spectrum herbicides) control or suppress plants regardless of species. For example, glyphosate (Roundup and others), glufosinate (Finale), and diquat (Reward) are nonselective herbicides.

Nonselective herbicides are often used to renovate turf areas or to trim along sidewalks and fences. Keep in mind that a particular herbicide may be labeled nonselective, but that does not mean it controls every weed. For example, glyphosate is a nonselective herbicide but it provides poor white clover control.

Preemergence and Postemergence Herbicides

Preemergence and postemergence refer to the time when the weed is controlled.

Preemergence herbicides are applied to a site before weed seeds germinate and prevent weeds from properly emerging or developing. Irrigation or rainfall activates preemergence herbicides.

As the name suggests, postemergence herbicides control weeds after they have emerged and are visible.

Herbicide Movement in Plants

Herbicides can be classified as systemic or contact.

Systemic herbicides move throughout the plant (they are translocated in plant vascular tissues). This movement of the herbicide within the plant is important for control because this allows the herbicide to move from a plant's leaves to its growing points and storage organs to help control them. Both selective and nonselective herbicides can be systemic. Most turf herbicides are systemic.

Contact herbicides do not move inside a plant. Instead, they affect only the plant tissues they directly contact. Contact herbicides are typically fast acting and include bromoxynil (Buctril), bentazon (Basagran T/O), and diquat (Reward).

Good spray coverage is essential when applying a contact herbicide and less critical (although still important) when using a systemic herbicide. For more information about herbicide movement, see *Herbicide Movement,* page 110.

Herbicide Mechanism of Action

Mode of action describes the plant processes that the herbicide affects and the sequence that leads to the death of susceptible plants. This includes absorbing, translocating, and metabolizing the herbicide.

The **mechanism of action** more specifically describes the specific biochemical site within a plant where the herbicide acts to inhibit or stimulate plant processes. For many mechanisms of action, the symptomology on plants is specific to the product, which can help applicators monitor the activity of the herbicide and detect off-target damage.

The Weed Science Society of America (WSSA) assigns numerical codes to the various herbicide mechanisms of action—*Herbicide Mechanism of Action*, page 110. For example, the code for synthetic auxin herbicides is WSSA Group 4. The following images show target weeds and off-target damage that illustrate the mechanisms of action of some herbicides with distinct symptomology. The photos include the WSSA code for each herbicide.

For a list of the mechanisms of action for the herbicides discussed in this publication, see *Common and Trade Names and Mechanism of Action of Registered Herbicides,* page 107.

White clover injury from the synthetic auxin herbicide dicamba (WSSA Group 4). Notice the bending and twisting of the stems.

Wild violet injury from the synthetic auxin herbicide triclopyr (WSSA Group 4). Notice the bending and twisting of the stems.

Injury to tomato from a synthetic auxin herbicide (WSSA Group 4). Notice the bending, twisting, and curling of the leaves.

Buckhorn plantain injury from the synthetic auxin herbicide quinclorac (WSSA Group 4). Notice the bending and twisting of the stems.

Off-target injury to white pine from a synthetic auxin herbicide (WSSA Group 4). Notice the bending and twisting of the stems.

Off-target injury to lilac from a synthetic auxin herbicide (WSSA Group 4). Notice the strapping symptom.

Ground ivy injury from the PPO inhibiting herbicide sulfentrazone (WSSA Group 14) tank-mixed with a synthetic auxin herbicide (WSSA Group 4). Tissue necrosis is rapid from sulfentrazone.

Dandelion injury from the synthetic auxin herbicide 2,4-D (WSSA Group 4). Notice the bending and twisting of the stems.

Despite the bending and twisting of the stems noticeable in this purple coneflower (Echinacea purpurea), this is not herbicide injury. Plants may naturally bend and twist due to environmental conditions or injury from other pests.

Bermudagrass injury from the mitotic inhibitor herbicide prodiamine (WSSA Group 3). Notice the swollen, "clubbed" root on the stolon.

Wheat injury from the HPPD herbicide mesotrione (WSSA Group 27). Notice the bleaching symptoms.

Smooth crabgrass injury from the ACCase inhibiting herbicide fenoxaprop (WSSA Group 1). Notice the necrotic base of the stem.

Dandelion injury from the HPPD herbicide mesotrione (WSSA Group 27). Weeds may take on a purple or pink color before turning white.

Dandelion injury from the ALS inhibitor herbicide florasulam (WSSA Group 2). Notice the reddening of the weed as it dies.

Yellow nutsedge injury from the ALS inhibiting herbicide halosulfuron (WSSA Group 2). Symptoms are chlorosis followed by necrosis.

Herbicide Families

Herbicides with similar chemical properties and activities are grouped into chemical families. For example, phenoxy-carboxylic acids (2,4-D, 2,4-DP [dichlorprop], mecoprop [MCPP], MCPA), pyridine-carboxylic acids (clopyralid, fluroxypyr, triclopyr), and benzoic acids (dicamba) are all distinct chemical families based on their chemical structures and activities, but all of these products have the same mechanism of action: synthetic auxins (WSSA Group 4).

Herbicide Resistance

Herbicide **resistance** is the inherited ability of a weed to survive and reproduce even though it was exposed to a normally lethal dose of herbicide.

Herbicide **tolerance** is the ability of a plant to remain uninjured by a dose of herbicide normally lethal to other plant species.

There are many more examples of herbicide-resistant weeds in other crops (such as corn or soybean) than in turf since the turf itself acts as a natural weed deterrent and because there are fewer weed species in mown turf than in agricultural fields. Typically, a weed may become resistant to a particular family of herbicides that have a unique mechanism of action.

For example, dinitroaniline-resistant goosegrass—the dinitroaniline chemical family (WSSA Group 3) includes prodiamine, pendimethalin, benefin, trifluralin, and oryzalin—has been documented in turfgrass; however, these resistant goosegrass populations are still susceptible to oxadiazon (Ronstar), which has a different mechanism of action (WSSA Group 14) and is an oxadiazole herbicide (oxadiazole chemical family). In the Midwest, there are few reports of weed resistance in turf systems, although smooth crabgrass is known to be resistant to quinclorac at some Indiana locations. And recently, populations of 2,4-D-resistant buckhorn plantain were discovered. Utilizing sound cultural practices to enhance turf growth, rotating herbicide mechanism of action, and applying herbicide mixtures are best practices to prevent herbicide resistance.

Herbicide Interactions and Mixing Order

When you mix two or more pesticides in a tank, the result is typically an additive effect—that is, each pesticide functions independently and produces a predictable result. However, mixing pesticides can cause certain interactions including:

- Antagonism
- Synergism
- Safening
- Chemical and physical incompatibility

Antagonism

Antagonism is the term that describes when one or more of the herbicides in a tank-mix provide reduced weed control.

The best example of this is mixing postemergence phenoxy broadleaf herbicides with the postemergence grass herbicides (graminicides) fenoxaprop (Acclaim Extra), fluazifop (Fusilade II), and sethoxydim (Segment II). If you mix a grass herbicide with 2,4-D or MCPA (or with herbicides that contain 2,4-D or MCPA such as ChangeUp, Trimec, Triplet, and others), annual grassy weed control will be reduced but broadleaf weed control may be unchanged. This antagonism is well-documented, especially with crabgrass control.

Therefore, it is always a good idea to avoid tank-mixing grass herbicides with 2,4-D or MCPA when attempting to control annual grassy weeds and broadleaf weeds. While we do not have data on all the herbicides that can be tank-mixed with grass herbicides without causing antagonism, we know that clopyralid, dicamba, fluroxypyr, and MCPP can be safely tank-mixed without causing antagonism. To avoid the risk of antagonism with 2,4-D or MCPA, apply grass herbicides separately (seven or more days apart if applying 2,4-D or MCPA first; two or more days apart if applying the grass herbicide first).

Poor water quality can also cause antagonism. See *Does the pH and hardness of my spray water influence weed control?*, page 64 for more information.

Synergism

Synergism occurs when the activity of the herbicide mixture is greater than the activity of the individual herbicides would have been if they were applied alone. There are few examples of true synergism of turf herbicides, although many provide beneficial additive affects when tank-mixed.

Safening

Although synergism is rare when tank-mixing turf herbicides, another positive effect that can occur when doing so is **safening**. Safening occurs when two ingredients (typically one injurious to turf and one noninjurious) are tank-mixed and the resulting application reduces the level of injury caused by the injurious herbicide.

The best examples of this in turf are when fenoxaprop (Acclaim Extra) or fluazifop (Fusilade II) are tank-mixed with triclopyr (Turflon Ester Ultra or Triclopyr 4) to suppress bermudagrass. Tank-mixtures of fenoxaprop + triclopyr or fluazifop + triclopyr reduce injury to desirable turfgrass species without reducing grassy weed control. Typically, these tank-mixtures are used to suppress bermudagrass in cool-season turf or in zoysiagrass (see *Bermudagrass*, page 43). In addition, you can reduce the bleaching effect of mesotrione (Tenacity) and topramezone (Pylex) by tank-mixing with triclopyr without reducing weed control. Another way to safen an herbicide is by adding an ingredient that is not an herbicide. Two examples include the herbicides Manuscript (pinoxaden) and Recognition (trifloxysulfuron). These herbicides already contain either cloquintocet-mexyl or metcamifen as safeners in their formulation so that no mixing is required by the applicator.

Chemical and Physical Incompatibility

Chemical incompatibility is when a tank-mix results in increased turf injury.

This can occur when tank-mixing certain insecticides and herbicides. Turf can be injured when organophosphate insecticides—acephate (Orthene), chlorpyrifos (Dursban), malathion (Malathion), or triclorfon (Dylox)—are tank-mixed with certain herbicides. These injuries occur because the insecticides can temporarily render turf unable to metabolize herbicides that normally cause little to no injury.

This interaction is most likely to occur when an ALS inhibitor (WSSA Group 2) or HPPDinhibtor (WSSA Group 27) mechanism of action herbicide follows an organophosphate application. Most of the ALS inhibitor herbicides labeled for turf are labeled for use only on warm-season turf with the exceptions of halosulfuron (ProSedge, SedgeHammer, others) and florasulam (Defendor) on all cool-season grasses and metsulfuron (Manor, Mansion, MSM) on Kentucky bluegrass.

Few herbicide labels contain precautionary statements about this potential interaction, but the Tenacity (mesotrione) label does state: "Do not apply an organophosphate or carbamate insecticide within seven days of a Tenacity application as turf injury may occur." Topramezone (Pylex) and imazapic (Plateau) labels contain similar statements. Carbaryl (Sevin) is a carbamate insecticide. To avoid these potential interactions, consider using a pyrethroid insecticide such as bifenthrin (Talstar) or cyfluthrin (Tempo).

Physical incompatibility can occur when two or more products are tank-mixed and form a solid substance that precipitates out of solution.

To prevent this interaction use a jar test. Detailed instructions on conducting a jar test are found in the Purdue Publication, *Avoid Tank Mixing Errors: A Guide to Applying the Principles of Compatibility and Mixing Sequence* (Purdue Extension publication PPP-122), which describes how several factors influence product compatibility. It is important to understand these basic factors (including formulation, tank-mixing order, and carrier type) to help you effectively combine crop protection products. The publication is available from the Education Store, https://edustore.purdue.edu/.

It is important to mix pesticide products in a specific order (both with the jar test and in the spray tank). Wear the appropriate PPE when mixing (see labels for requirements). Unless label directions specify a different mixing order, follow this general order:

1. Read all product labels.
2. Shake all liquid product containers.
3. Fill the spray tank with 50%of the required water volume.
4. Start the agitation and continue through the mixing process.
5. Add products based on formulation type in the order listed here:
   a. Water-soluble packets (WSP) –proceed to step 5b only after packets fully dissolve.
   b. Dry formulations –such as wettable powders (WP), water dispersible granules (WDG), and dry flowables (DF)
   c. Ammonium sulfate (dry or liquid)
   d. Compatibility agents and defoamers (if needed) –wait 2–3 minutes before adding other products
   e. Dispersed liquid formulations –such as flowables (F) and suspension concentrates (SC), suspoemulsions (SE), emulsifiable in water (EW), and micro-encapsulated (ME)
   f. Liquid drift retardants (if desired)
   g. Remaining liquid formulations –such as emulsifiable concentrates (EC), oil dispersions (OD), and solutions and soluble liquids (S or SL)
   h. Adjuvants –such as crop oil concentrates (COC), methylated seed oil (MSO), nonionic surfactants (NIS), water conditioning agents, and other adjuvants not listed above
   i. Micronutrients and liquid fertilizers
6. Wait 3 to 5 minutes before adding products after dry formulations (step 5b).
7. Add remaining water.

Measure the pH of your spray solution. Only add pH adjusters at the end of the mixing process to ensure that the final spray solution is in the pH range specified by the product label.

Sprayer and Spreader Calibration

There is no substitute for properly calibrated equipment. Precise application allows the herbicide to function as it was intended. Improper calibration may cause poor weed control if the application rate is too low, or it may injure the desired turf if the rate is too high.

Maybe even more importantly, inadequate applications cost you money—in labor, equipment, and fuel—because you have to make repeated applications. Overapplication also is costly—weed control products can be expensive. And when you operate improperly calibrated equipment, customers and clients may question your professionalism and knowledge. When weed problems persist, customers wonder why they are paying you.

Find Out More

Purdue Extension offers various publications about calibration:

- *Calibrating the Hose Reel Lawn Care Sprayer* (PPP-85)
- *Calibrating Ride-on Pesticide Sprayers and Spreaders: Keys to Application Accuracy* (PPP-104)
- *Lawncare Pesticide Application Equipment: A Guide to Selection and Calibration of Granular Spreaders* (PPP-46)
- *Category 3b: Turf Pest Management* (PPP-3B) and six other training manuals cover the calibration of sprayers

Purdue Extension publications are available from the Education Store:

https://edustore.purdue.edu/ • (765) 494-6794

Why Didn't the Herbicide Kill the Weed?

There are many possible reasons an herbicide doesn't control a weed, but the most common is human error.

Here are some things to consider when trying to determine why an figure herbicides didn't work:

- Did you correctly identify the weed? See pages 5–35 for weed identification help.
- Did the herbicide label list the weed you wanted to control?
- Did you use the appropriate application rate on the product label?
- Did you achieve uniform spray coverage during the application?
- Did you apply the herbicide to the weed when it was small or large? Larger, more mature weeds are more difficult to control. See product labels for information about control at weed growth stages.
- Was the temperature during the application within the label's recommendations?
- Was the soil moisture adequate for proper weed growth and herbicide uptake?
- Did you tank-mix herbicides that were antagonistic?
- Did the label recommend two applications (as they sometimes do with difficult-to-control weeds)?
- If you applied a postemergence herbicide, did rain or irrigation wash it off within a few hours after application?

- Did you include the adjuvant or surfactant recommended on the label?
- If you applied a preemergence herbicide, did you water it within a few days of application?
- If you applied a preemergence herbicide, had the weed already germinated and emerged when the application was made?

Control of Tough Weeds

Weeds can be difficult to control for many reasons, including their ability to spread, to produce seed, and to tolerate herbicides, as well as their similar biology to many desirable turfgrasses. As such, a more specific approach or strategy may be required to achieve control. This section provides site-specific recommendations for difficult-to-control turf weeds.

Annual Bluegrass

Annual bluegrass (*Poa annua*—see photos on page 27) is possibly the toughest weed to control in turfgrass because its biology is so similar to desirable turfgrass species and it produces a prolific number of seeds. Annual bluegrass invades turf especially when mowing heights are less than 2 inches, irrigation is frequent (or drainage is poor), and where shade is present. While this plant is classified as a winter annual, weak perennial biotypes exist. Because of its life cycle, it is very noticeable in late spring (when flowering) and summer (when it declines during heat, disease, and drought stress).

Annual bluegrass control recommendations are specific to the turfgrass species and use area you are treating. Cultural controls include raising the mowing height, reducing irrigation, reducing shade, alleviating soil compaction, removing grass clippings during flowering, and decreasing fertilization.

One strategy is to prevent annual bluegrass from invading areas by using a preemergence herbicide. Preemergence herbicides—such as bensulide (Bensumec), dithiopyr (Dimension), pendimethalin (Pendulum), prodiamine (Barricade), sulfentrazone + prodiamine (Echelon), and others—can be applied in late summer to prevent this winter annual from germinating in turf after a stressful summer.

Following the late summer application before annual bluegrass germination, making a sequential late-fall application of the preemergence herbicide will help to prevent annual bluegrass encroachment. This preemergence control strategy will be ineffective if a cool, moist summer allows annual bluegrass to persist or where weak perennial biotypes are present. A disadvantage of this control strategy is that applying preemergence herbicides in late summer also prevents seeding desirable turf species in late summer and fall. If this is the situation, there are postemergence herbicides available to control annual bluegrass.

Different turf areas have different postemergence control options. Here are some tips for controlling annual bluegrass in the following turf areas:

- creeping bentgrass fairways.
- creeping bentgrass putting greens.
- Kentucky bluegrass.
- perennial ryegrass and tall fescue.
- bermudagrass and zoysiagrass.

In Creeping Bentgrass Fairways

Plant growth regulators (PGRs) such as paclobutrazol (Trimmit), paclobutrazol + trinexapac-ethyl (Legacy), or flurprimidol (Cutless) are effective in managing annual bluegrass in creeping bentgrass fairways. These products are used in programs designed to gradually reduce annual bluegrass. This approach requires patience, consistency, and a fairway management program designed to favor the creeping bentgrass. In most cases, especially where annual bluegrass is well-established, these herbicides will not provide complete annual bluegrass control after only a single year of use.

Xonerate 2SC (amicarbazone) may injure creeping bentgrass that is stressed (1) when applied in the summer or fall, (2) when applied to juvenile turf, or (3) when turf is shallow-rooted. Apply Xonerate 2SC in the spring when daily high temperatures are between 50°F and 80°F and turf is actively growing. Apply Xonerate 2SC at 3.0 fl oz/A at 14-day intervals for a maximum of four applications or at 6.0 fl oz/A at 14-day intervals for a maximum of two applications.

Prograss (ethofumesate) is most effective when applied in fall as two sequential applications spaced three to four weeks apart. A spring Prograss application may also improve control when applied following fall applications. Prograss may injure certain older creeping bentgrass cultivars when applied as indicated on the label.

Velocity PM (Poa Management) controls both *Poa annua* and *Poa trivialis* and injures many *Poa pratensis* (Kentucky bluegrass) cultivars. Make sure to assess how much of each of these species is in the fairway before using this herbicide. If more than 10% of the fairway turf is *Poa* spp., consider use carefully. Application timing for Velocity PM is between late spring and early fall when daytime highs are between 70°F and 85°F. Make two applications of Velocity PM at 2.25 fl oz/A at 14-day spray intervals to creeping bentgrass as a more aggressive control approach, or use the less aggressive approach of up to five applications of Velocity PM at 0.38 to 0.75 fl oz/A at 3- to 7-day spray intervals. As always, read the complete label instructions before making an application.

In Creeping Bentgrass Putting Greens

Annual bluegrass is the most troublesome weed in golf course putting greens. It tolerates low mowing and compaction, and thrives in the moist conditions found on putting greens. There is no single herbicide or plant growth regulator that can completely and safely control annual bluegrass in creeping bentgrass, but there are several options available to help golf course superintendents manage annual bluegrass.

The traditional approach is to use preemergence herbicides: dithiopyr, bensulide (Bensumec), or bensulide + oxadiazon (Goosegrass/Crabgrass Control). But this approach has the shortcoming of failing to control the perennial type of annual bluegrass (*Poa annua* var. *reptans*). Fall and spring applications over the course of many consecutive years are needed to suppress and control annual types (*Poa annua* var. *annua*).

An option that has shown promise is using plant growth regulators (PGRs) such as paclobutrazol (Trimmit), paclobutrazol + trinexapac-ethyl (Legacy), or flurprimidol (Cutless). These products are used in conversion programs that are designed to gradually reduce annual bluegrass. This approach requires patience, consistency, and a management program designed to favor the desirable turfgrass. Research at Purdue, Michigan State, and Nebraska has found that response to these products varies by location, indicating that some users may find them effective while others may not achieve desirable results.

Moghu Research Center in Korea developed an herbicide called methiozolin (PoaCure). It effectively controls annual bluegrass in creeping bentgrass putting greens and fairways, and it is a new control option for this tough-to-control weed.

In Kentucky Bluegrass

To control annual bluegrass, you can apply ethofumesate (Prograss) to all general turf areas with Kentucky bluegrass, including golf courses, sod farms, and commercial and residential turf. Annual bluegrass control with ethofumesate is most effective when applied in fall as two sequential applications spaced three to four weeks apart. A spring application of Prograss may also improve control when applied following fall applications. The Prograss application rate is lower for Kentucky bluegrass than other turfgrass species, which reduces the annual bluegrass control provided by this herbicide. Further, Prograss may injure certain Kentucky bluegrass cultivars when applied as indicated on the label.

Postemergence applications of mesotrione (Tenacity) in late fall injure annual bluegrass, but reports on annual bluegrass control in Kentucky bluegrass with Tenacity are inconsistent. Tenacity is labeled for preemergence suppression of annual bluegrass but not postemergence control.

For annual bluegrass control in Kentucky bluegrass, apply Xonerate 2SC at 3.0 fl oz/A at 14-day intervals for a maximum of four applications or at 6.0 fl oz/A at 14-day intervals for a maximum of two applications. Tank-mixing with common broadleaf herbicides will enhance control of broadleaf weeds. Make applications of Xonerate 2SC in the spring when turf is actively growing and daily high temperatures do not exceed 80°F. Do not apply in the summer or fall. Like other herbicides, annual bluegrass control with Xonerate is inconsistent, especially on perennial biotypes.

PGRs have some success when used in creeping bentgrass as part of an annual program. However, this approach is less likely to work in Kentucky bluegrass, perennial ryegrass, or fescues because PGRs that regulate annual bluegrass growth will also regulate these desirable turf species. In addition, most annual bluegrass in taller mown turf is a true winter annual type that dies each summer and germinates again each fall. Thus, regulating existing annual bluegrass would have no effect on the germination of annual bluegrass from seed.

In Perennial Ryegrass and Tall Fescue

Bispyribac-sodium (Velocity PM), amicarbazone (Xonerate 2SC), and ethofumesate (Prograss) are three herbicides registered for annual bluegrass control in perennial ryegrass and tall fescue. Velocity PM may injure perennial ryegrass that is stressed or not well established. Application timing for Velocity PM is between late spring and early fall when daytime highs are between 70°F and 85°F. Make up to three applications of Velocity PM at 0.75 to 1.1 fl oz/A at 21- to 28-day spray intervals to perennial ryegrass and tall fescue. Make monthly applications at 0.38 to 0.75 fl oz/A if a slower conversion is desired. As always, read the complete label instructions before making an application.

For annual bluegrass control in perennial ryegrass, fine fescues, and tall fescue, apply Xonerate 2SC at 6.0 fl oz/A at 14- to 21-day intervals for a total of two applications. Make applications of Xonerate 2SC in the spring when turf is actively growing and daily high temperatures do not exceed 85°F. Do not apply in the summer or fall.

You can apply Prograss to all general turf areas with perennial ryegrass and tall fescue for annual bluegrass control. It is most effective when applied in fall as two sequential applications spaced three to four weeks apart. A spring Prograss application may also improve control when it follows a fall application. Perennial ryegrass is generally tolerant to Prograss and should not be injured by it when applied at 0.66–1.33 gal/A (Prograss EC) or at 1–2 qt/A (Prograss SC) as indicated on the label. Tall fescue is generally tolerant to Prograss and should not be injured when applied at 0.5–1.0 gal/A (Prograss EC) or at 0.75–1.5 qt/A (Prograss SC) as indicated on the label. Because higher Prograss rates can be used on perennial ryegrass, annual bluegrass control from Prograss is often more successful in this species.

In Bermudagrass and Zoysiagrass

Annual bluegrass control in warm-season turf is more easily accomplished beause many herbicides that effectively control annual bluegrass have excellent safety on warm-season turfgrasses such as zoysiagrass and bermudagrass. Additional preemergence herbicides that are unsafe on cool-season turf—such as atrazine, dimethenamid (Tower), flumioxazin (SureGuard), indaziflam (Specticle), metolachlor (Pennant MAGNUM), oryzalin (Surflan AS), and simazine—are options on some warm-season turfgrasses.

Many sulfonylurea herbicides are labeled for the selective postemergence removal of annual bluegrass from bermudagrass and zoysiagrass. You can safely use these products in dormant bermudagrass and zoysiagrass or during spring green-up when bermudagrass is emerging from winter dormancy. Flazasulfuron (Katana), foramsulfuron (Revolver), rimsulfuron (Rimsulfuron 25DF), sulfosulfuron (Certainty), and trifloxysulfuron (Monument, Recognition) can all be used for annual bluegrass removal in bermudagrass and zoysiagrass. See *How can I safely remove overseeded turf (that is, perennial ryegrass) without injuring the bermudagrass?*, page 65 for more information.

Additionally, a nonselective herbicide may be used during winter to kill winter weeds without injuring the bermudagrass or zoysiagrass. See *Can I safely spray bermudagrass and zoysiagrass with a nonselective herbicide such as glyphosate to control weeds during winter?*, page 65 for more information. A new option is to apply flumioxazin (SureGuard) to dormant bermudagrass and zoysiagrass in late fall for postemergence and preemergence control.

Bermudagrass

While bermudagrass (*Cynodon* spp.—see photos on page 22) is a commonly used turfgrass in the transition zone and southern United States, it is also a difficult-to-control weed in northern turf.

Different situations require different control methods. The areas below describe bermudagrass:

- Control during renovation.
- Encroachment into creeping bentgrass putting greens.
- Selective suppression in other turf species.

Control During Renovation

For most, fumigation with dazomet (Basamid) or methyl bromide is not an option because it is so costly and completion of a special fumigation training program is required. A good alternative is to make three glyphosate applications (up to 3 lb ae/gal) at 3 qt/A over the growing season (June, July, and August). Wait three to four weeks for regrowth before making the follow-up applications.

Research shows that a tank-mix of 3 qt/A glyphosate + 24 fl oz/A of fluazifop (Fusilade II) will improve bermudagrass control over glyphosate alone. When using Fusilade II, be aware that it will have some residual soil activity. Wait at least 30 days before seeding turfgrass after applying Fusilade II to bare ground or 14 days after an application to turf.

Encroachment into Creeping Bentgrass Putting Greens

An encroachment management program should include frequent mechanical edging of the putting green and resodding of infested areas. Ethofumesate (Prograss) could be used to slow bermudagrass encroachment in the collars of bentgrass putting greens. However, this herbicide will cause varying degrees of bentgrass injury. You should make applications in the spring and/or fall when bentgrass is actively growing and bermudagrass is less active. Suppression is temporary (up to three months) with these treatments.

Tank-mixes of flurprimidol (Cutless) with Tupersan have sometimes given better results than Tupersan alone. Tupersan 50WP is used in spring or fall at 1 lb/1,000 ft^2. Repeat as needed and water-in after application. Temporary turf discoloration may result. Use a spray hawk to make applications to collars. Alternatively, you can apply a granular formulation of Tupersan with a drop spreader to your putting green collars to suppress bermudagrass encroachment. Another option is to use topramezone (Pylex) (see *Goosegrass*, page 50).

Selective Suppression in Other Turf Species

In zoysiagrass, use fluazifop (Fusilade II) at 3–4 fl oz/A or fenoxaprop (Acclaim Extra) at 28 oz/A to suppress bermudagrass. Adding triclopyr (Turflon Ester Ultra or Triclopyr 4) at 1 qt/A to either fluazifop or fenoxaprop will improve turfgrass safety and bermudagrass control. You can also tank-mix Recognition herbicide (trifloxysulfuron + metcamifen) with Fusilade II (fluazifop) for bermudagrass control in zoysiagrass. The metcamifen helps protect the zoysiagrass from herbicide injury. Begin around June 1 and repeat every four weeks unless the zoysiagrass has not recovered from the first application. Zoysiagrass growing in the shade is more susceptible to injury. To avoid antagonism, do not tank-mix fluazifop or fenoxaprop with phenoxy herbicides. Late fall applications to zoysiagrass may cause increased injury.

In tall fescue, repeat applications of fenoxaprop (Acclaim Extra) at 28 fl oz/A or fluazifop (Fusilade II) at 5–6 fl oz/A at 28- to 35-day intervals during the late spring and late-summer/fall provide bermudagrass suppression. Adding triclopyr (Turflon Ester Ultra or Triclopyr 4) at 1 qt/A to either fluazifop or fenoxaprop will improve turfgrass safety and bermudagrass control. Applications in late spring (May) when bermudagrass is emerging from winter dormancy and in late-summer/fall (September/October) prior to winter dormancy are recommended for best results and increased safety on tall fescue. To avoid antagonism, do not tank-mix with phenoxy herbicides. Do not use on tall fescue seedlings less than four weeks old.

An alternative to Acclaim Extra is the fenoxaprop-containing herbicide, Last Call (see its label for instructions regarding bermudagrass control). Another option is to make three applications of topramezone (Pylex) at 1.33 fl oz/A at 21-day intervals starting in late summer to suppress bermudagrass. For best results, tank-mix topramezone with triclopyr at 1 qt/A and include a methylated seed oil at 0.5%–1.0% (v/v; that is, 0.5 to 1.0 gal MSO volume per 100 gal water).

In perennial ryegrass and Kentucky bluegrass, repeat applications of fenoxaprop (Acclaim Extra) at 28 oz/A at 28- to 35-day intervals during the late spring and late summer/fall provide bermudagrass suppression. Adding triclopyr (Turflon Ester Ultra or Triclopyr 4) at 1 qt/A to Acclaim Extra will improve turfgrass safety and bermudagrass control. Applications in late spring (May) when bermudagrass is emerging from winter dormancy and in late summer/fall (September/October) prior to winter dormancy are recommended for best results and increased safety on perennial ryegrass and Kentucky bluegrass. To avoid antagonism, do not tank-mix with phenoxy herbicides. Do not use on turf seedlings less than four weeks old. An alternative to Acclaim Extra is the fenoxaprop-contianining herbicide Last Call (see its label for instructions regarding bermudagrass control). Another option is to make three applications of topramezone (Pylex) at 1.33 fl oz/A at 21- to 28-day intervals starting in late summer to suppress bermudagrass. For best results, tank-mix topramezone with triclopyr at 1 qt/A and include a methylated seed oil at 0.5%–1.0% (v/v; that is, 0.5 to 1.0 gal MSO volume per 100 gal water).

Broadleaf Plantain—*see Buckhorn Plantain*

Buckhorn Plantain

Buckhorn plantain (*Plantago lanceolata*—see photos on page 20) and broadleaf plantain (*Plantago major*—see photos on page 20) are common weeds in managed and unmanaged turf. Both are perennials. Buckhorn plantain is more common in unirrigated, thin turf that is mown infrequently. Broadleaf plantain is more common in irrigated, managed areas. However, both can survive tough environmental conditions in a wide range of soils.

Many herbicides effectively control plantains, especially those that contain 2,4-D. Fall is the best time to control perennial broadleaves like plantains, and both amine and ester 2,4-D formulations provide optimum control in the fall. To learn which herbicides control plantains best, see *Weed Control Ratings for Postemergence Broadleaf Herbicides*, page 76.

Generally, herbicides that control buckhorn plantain will also control broadleaf plantain; however, there are a few exceptions.

First, quinclorac (Drive XLR8 and others) is fair for buckhorn plantain control but poor for broadleaf plantain.

Second, when using herbicides containing clopyralid (Lontrel, Confront, and others), effective broadleaf plantain control usually requires the highest label rate.

Finally, buckhorn plantain response to 2,4-D is quite variable from one population to another; 2,4-D easily controls some populations (especially when applied in the fall), while other populations are resistant to it. If you suspect 2,4-D resistance in buckhorn plantain, clopyralid or halauxifen-methyl-containing herbicides provide excellent control.

Bull Thistle—*see Thistles*

Canada Thistle—*see Thistles*

Common Dandelion—*see Dandelion*

Common Plantain—*see Buckhorn Plantain*

Common Purslane—*see Purslane*

Crabgrass

Large crabgrass (*Digitaria sanguinalis*—see photos on page 23), smooth crabgrass (*Digitaria ischaemum*—see photos on page 23), and to a lesser extent southern crabgrass (*Digitaria ciliaris*) are all species found in the Midwest that are collectively referred to as crabgrass. Crabgrass is often considered to be the most problematic weed in lawns. It is a summer annual grass that typically germinates in April in the Midwest and Northeast (late March in southern areas and late April in northern areas).

Once conditions are favorable for germination, crabgrass can continue to germinate throughout the summer until August. One square foot of crabgrass can produce at least 10,000 seeds, so crabgrass becomes an annual problem in most turf areas. Proper mowing (higher mowing heights), proper fertilization (some rather than none to improve turf density), irrigation to prevent summer dormancy during drought, and aerification of compacted areas to improve turf health are all cultural practices that can be used to reduce crabgrass.

Crabgrass can be controlled using preemergence or postemergence products.

Crabgrass Development Stages

Knowing the size or stage of development of annual grasses like crabgrass is important for selecting the proper herbicide for control. Herbicide labels contain notes that describe how well they work on certain leaf or tiller sizes of annual grasses.

Smaller annual grasses (pre-tiller) in the 1-5 leaf stage are easier to control than larger, tillered grasses. In most cases, tillered annual grasses like crabgrass require two postemergence applications for complete control.

The developmental stages for large crabgrass are shown on the following pages, but other crabgrasses, goosegrass, foxtails, barnyardgrass, and sandbur developmental stages can also be identified using similar terminology.

One-leaf stage.

Two-leaf stage.

Three-leaf stage.

Four-leaf stage.

Five-leaf stage.

One-tiller stage.

Two-tiller stage.

Three-tiller stage.

Four-tiller stage.

Five-tiller stage.

Eight-tiller stage.

Seedhead stage.

Preemergence Control of Crabgrass

Despite proper cultural practices, crabgrass may still remain problematic in certain hot spots such as next to sidewalks and driveways and in sunny turf areas. The best approach to controlling crabgrass is to use a preemergence herbicide such as dithiopyr (Dimension), pendimethalin (Pendulum), prodiamine (Barricade), prodiamine + quinclorac (Cavalcade PQ), sulfentrazone + prodiamine (Echelon), and others. These herbicides inhibit cell division and prevent crabgrass seeds from properly emerging. Since these herbicides work on germinating seeds, you must apply them prior to germination—with the exception of dithiopyr, which controls crabgrass after germination until it reaches one tiller.

Here are some tips for controlling crabgrass with preemergence herbicides:

- Dithiopyr, pendimethalin, and prodiamine-containing herbicides provide similar crabgrass control in the Midwest and Northeast. Typically, these ingredients also provide better crabgrass control than older chemistries such as benefin (Balan), benefin + trifluralin (Team Pro), or bensulide (Bensumec).
- Other preemergence herbicides work well on crabgrass but they are labeled for use on warm-season turf only. These include benefin + oryzalin (Surflan XL), dimethenamid (Tower), dimethenamid + pendimethalin (Freehand), indaziflam (Specticle FLO or Specticle G), and oryzalin (Surflan).
- Preemergence herbicides require irrigation or rainfall after application to be activated and for best results.
- A common mistake is to use too low of an application rate. Consult the label for the exact rates for the formulation. In the Midwest and Northeast, we typically recommend a rate of 0.5 lb ai/A of dithiopyr (emulsifiable or WSP), 0.38 lb ai/A of dithiopyr G (granular), 3.0 lb ai/A of pendimethalin (liquid or dry), or 0.75 lb ai/A of prodiamine (liquid or dry).
- Split or sequential applications are a strategy for improving crabgrass control with preemergence herbicides. The general strategy is to split one application into two and use the same total amount of product. The first application is made at the normal timing before crabgrass germinates at half the normal rate; the second half of the application is made 45–60 days later. For example, the normal rate for Dimension 2EW is 2 pt/A or 0.5 lb ai/A in April. For a split application, apply 1 pt/A (0.25 lb ai/A) on April 1, and then apply 1 pt/A (0.25 lb ai/A) about 60 days later on June 1.

Postemergence Control of Crabgrass

Another approach to controlling crabgrass is to use postemergence herbicides after crabgrass has already emerged.

Here are some tips for controlling crabgrass with postemergence herbicides:

- Know what stage (size) of crabgrass is present before you select and apply the herbicide. Herbicide labels contain notes that describe how well they work on certain leaf or tiller sizes of crabgrass. Learn how to accurately identify the size of the crabgrass in *Crabgrass Development Stages*, page 45.
- Smaller crabgrass (pre-tiller) is easier to control than larger, tillered crabgrass. In most cases, tillered crabgrass requires a higher application rate to control or two postemergence applications made at a two- to three-week interval for complete control. The *Effectiveness of Postemergence Herbicides on Crabgrass at Various Growth Stages* on page 48 table summarizes the label statements about postemergence crabgrass control.
- Most postemergence herbicides do not provide preemergence control. So applications in June or early July may kill small plants, but new seedlings often emerge in their place. Use a combination of postemergence and preemergence herbicides for control in late spring where infestations are severe or turf cover is poor.
- Dithiopyr can provide effective postemergence crabgrass control on plants that have up to one tiller, in addition to providing preemergence control.
- MSMA can effectively control postemergence crabgrass but is only labeled for use on golf courses and sod farms. Read the label for full instructions.
- Quinclorac can be used on most cool- and warm-season grasses. It effectively controls crabgrass when crabgrass is one tiller or smaller or when it is has more than four tillers. The reason for this is unclear but well documented. Products that contain quinclorac include Cavalcade PQ, Drive XLR8, Quinclorac 75DF, Quinclorac 1.5L, Momentum Q, Q4 Plus, Quincept, and Solitare.
- Mesotrione (Tenacity) is also effective for postemergence crabgrass control. In most cases, it will be necessary to make two applications at 5 oz/A at a two-week interval to effectively control crabgrass. Similarly, applications of topramezone (Pylex) at a three-week interval will control crabgrass. Research has demonstrated that control from these herbicides is increased when nitrogen fertilizer (0.125 to 0.5 lb N/1,000 ft^2) is applied a few days before or the day of an application.
- Fenoxaprop (Acclaim Extra) and Last Call (fenoxaprop + fluoroxypr + dicamba) are also effective at controlling crabgrass. Do not tank-mix Acclaim Extra or Last Call with herbicides that contain 2,4-D. See *Antagonism,* page 40.
- Applying postemergence herbicides at temperatures greater than 85°F will increase the risk of turf injury. In weedy turf, some turf injury might be acceptable in order to control weeds, but in turf areas with relatively few weeds it would be best not to make a postemergence herbicide application for crabgrass control when daytime highs are expected to exceed 85°F.

Effectiveness of Postemergence Herbicides on Crabgrass at Various Growth Stages

Herbicide[1]	1-leaf	2-leaf	3-leaf	4-leaf	5-leaf	1-tiller	2-tiller	3-tiller	4-tiller	≥5-tiller	6-tiller
dithiopyr (Dimension and others)	Apply before tillering.[2]										
fenoxaprop (Acclaim Extra and Last Call)	The label provides instructions for postemergence applications to annual grasses up to 5-tiller. The label advises which rate to use based on the size of the annual grassy weed.										3
mesotrione (Tenacity and others)	The label advises that postemergence applications be made to grasses before 4-tiller. Efficacy is reduced as plant size increases.								4		
MSMA[5]	Best results obtained on young, actively growing weeds										
quinclorac (Drive and others)	Growth stage for optimum contro.l						Applications may not provide complete control (2- to 4-tiller).			Growth stage for optimum control	
topramezone (Pylex)	Use higher rates on crabgrass >1-tiller.[7]									7	

1 Information based on herbicide labels and/or research experience. Follow label instructions regarding safety on desired turfgrass species, application rate, appropriate adjuvants to use with each product, and whether the product can be safely tank-mixed with other herbicides.

2 Dithiopyr is for use on crabgrass prior to tillering.

3 Acclaim Extra label instructions are not supplied for annual grasses >5-tiller. Use 4 pt/A of Last Call when treating annual grasses with >2 tillers. As plant size increases, efficacy decreases for both herbicides.

4 Tenacity label instructions advise applications to "<4-tiller crabgrass and goosegrass for best results." Efficacy is reduced as plant size increases.

5 MSMA is labeled for use on golf courses and sod farms. Read the label for full instructions. Best results obtained on young actively growing weeds.

6 Drive and quinclorac-containing herbicide labels advise that applications "made to annual grasses at 2- to 4-tiller may not provide complete control." Optimum control achieved when plants are <2-tiller or >4-tiller ("as weed grasses mature"). Sequential applications will be required when treating 2- to 4-tiller crabgrass.

7 Pylex label instructions advise 1.0 fl oz/A when annual grasses are <2-tiller in size, 1.5 fl oz/A when annual grasses are 2- to 4-tiller in size. Efficacy is reduced as plant size increases.

Creeping Bentgrass

Creeping bentgrass (*Agrostis stolonifera*—see photos on page 21) is a problematic weed on golf courses, athletic fields, and lawns. Creeping bentgrass contamination in stands of other cool-season species is common because there are limited cultural control and selective herbicide control options.

Traditionally, turf managers have attempted to control spreading grasses like creeping bentgrass with a nonselective systemic herbicide such as glyphosate (Roundup). Best results are seen when the weedy plants are young, fully green, actively growing, and not under drought stress. It is important to allow the weed to regrow before the next application. At least two applications are recommended, but three or more may be needed. This will leave the area dead and unsightly for a number of weeks or months if you desire optimum control.

The selective herbicide mesotrione (Tenacity) is labeled for use in turf to control a wide spectrum of weeds including annual grassy weeds, broadleaf weeds, yellow nutsedge, and tough perennial grassy weeds such as creeping bentgrass and nimblewill. Mesotrione is safe for use in Kentucky bluegrass (5–8 fl oz/A), perennial ryegrass (5 fl oz/A), tall fescue (5–8 fl oz/A), and fine fescue (5 fl oz/A). Among these species, Kentucky bluegrass is the most tolerant. Injury can be seen on tall fescue, perennial ryegrass, and fine fescue when the labeled use rate is exceeded or during certain environmental conditions. Mesotrione also injures annual bluegrass.

Mesotrione provides selective control of creeping bentgrass in cool-season turf with multiple applications made at the appropriate timing. Keep these factors in mind when controlling creeping bentgrass with mesotrione:

- **Timing.** Begin applying mesotrione in the fall (early September in the Midwest). Spring and summer applications provide less consistent control. Mesotrione loses effectiveness on creeping bentgrass later in the fall so it is important to make your first application no later than the second week of October (based on West Lafayette, Indiana, data).
- **Sequential applications.** You will need to make multiple mesotrione applications to control creeping bentgrass. Single applications are largely ineffective. Make three or four mesotrione applications at two-week intervals. Intervals of more than two weeks will reduce efficacy. At least three applications are required for best results.
- **Rate.** If you make three applications, use the 5 fl oz/A rate (or 6, 6, and 4 fl oz/A). If you make four applications, use the 4 fl oz/A rate. The product does have a yearly maximum use rate and the label states "do not apply more than 16 oz of Tenacity per acre per year or per crop (equivalent to a maximum of 0.50 lb of mesotrione per acre per year)."
- **Mowing height.** Typically, it is easier to control creeping bentgrass that has been mown at a lower height.
- **Adjuvant.** Adding urea ammonium nitrate (UAN) at 2.5% (v/v; that is, 2.5 gal UAN volume per 100 gal water) to the spray tank will improve creeping bentgrass control.
- **Surfactant.** The label recommends adding a nonionic surfactant (NIS) at 0.25% (v/v; that is, 0.25 gal (1 qt) NIS volume per 100 gal water) for best results.
- **Cultural practices.** You can improve control after the first mesotrione application by raking the creeping bentgrass to remove dead leaf tissue.

- **Adjustments.** Reducing the rate with more frequent application intervals can also effectively control creeping bentgrass. Rates as low 2 oz/A applied six times on a five-day interval provided equivalent control to three applications of 5 oz/A on a two-week interval.

Although mesotrione is the best selective herbicide we have available for creeping bentgrass control, it cannot completely control creeping bentgrass with just one season of applications. Using the appropriate timings, applications, rates, surfactants, adjuvants, and cultural practices, three mesotrione applications will provide greater than 90% control of creeping bentgrass. Any creeping bentgrass that was not killed one season will return the next and spread. Therefore, multiple applications over a number of years will be needed for complete creeping bentgrass control with mesotrione.

Creeping Charlie—*see Ground Ivy*

Dallisgrass

Dallisgrass (*Paspalum dilatatum*—see photos on page 26), thin Paspalum (*Paspalum setaceum*), and field paspalum (*Paspalum laeve*—see photos on page 26) are very similar warm-season perennial grassy weeds with short rhizomes. These species are among our most difficult-to-control turf weeds. Although these species have short rhizomes, they effectively are a bunch-type plant and form scattered clumps of grass in the landscape. They can also spread from one area to another by seed. The most effective way to control these weeds is to dig up the clumps in the turf with a shovel.

Another option is spot treatment with the nonselective herbicide glyphosate (Roundup and others). At least two glyphosate applications are needed. Apply when the plant first greens up in the spring (late April or May), and again when regrowth appears. Multiple follow-up applications may be required. You can also spray dallisgrass and field paspalum in October before they turn off color and enter winter dormancy. Use a 1%–2% spray solution (1.3–2.6 fl oz of glyphosate/gal of water).

Obviously, glyphosate is going to kill some of the desirable grass and could leave big brown spots in the turf so spot apply the herbicide only to the weed and not to the turf, if possible. You can even use a paintbrush, sponge, or foam applicator to apply the glyphosate only to the weed to help reduce the risk of injuring the desirable turf.

In Kentucky bluegrass, fine fescue, tall fescue, or perennial ryegrass lawns, three applications of mesotrione (Tenacity) at 5 oz/A spaced with a two-week application interval should reduce these paspalum species about 50%. Topramezone (Pylex) is more effective on dallisgrass than mesotrione but still only provides partial control. Make three applications at 1.33 fl oz/A at 21-day intervals. Start making these applications in June.

In tall fescue, two applications of fluazifop (Fusilade II) at 5–6 fl oz/A as spot treatments at three- to four-week intervals starting in late April and September will suppress field paspalum, thin paspalum, and dallisgrass.

Similar to tall fescue (but at a higher rate), two applications of fluazifop (Fusilade II) at 8–16 fl oz/A as spot treatments will suppress paspalum species in fine fescues.

In bermudagrass, three or more MSMA applications at one- to two-week intervals are necessary, with applications starting in May. In some studies, applying MSMA, waiting two weeks, and then spot treating with foramsulfuron (Revolver), sulfosulfuron (Certainty), or trifloxysulfuron (Monument, Recognition), followed by a second MSMA application one week later has, improved control.

Note: MSMA cannot be purchased as of December 31, 2010 (except for golf courses and sod farms), but applicators may use their remaining supplies.

Manuscript (pinoxaden) spot treatments at 19.2 fl oz/A applied once or two applications at 9.6 fl oz/A at a two-week interval in late summer will suppress many *Paspalum* spp. in annual bluegrass, bermudagrass, fine fescues, and zoysiagrass turf.

Tribute TOTAL (thiencarbazone + foramsulfuron + halosulfuron) at 3.2 oz/A also is effective at controlling dallisgrass in bermudagrass and zoysiagrass turf when making multiple applications in later summer and early fall before winter dormancy and a subsequent follow-up application in spring.

In zoysiagrass, two applications of fluazifop (Fusilade II) at 3–4 oz/A at three- to four-week intervals during the summer will suppress *Paspalum* spp.

Dandelion

Dandelion (*Taraxacum officinale*—see photos on page 49) is the most easily identified weed in lawns and possibly the most common. Dandelion is a perennial broadleaf weed that forms a deep taproot that helps it survive tough environmental growing conditions in a wide range of soils, although it does prefer soils high in potassium. It is a prolific seed producer and the seed readily germinates, which helps keep this weed problematic.

Many herbicides effectively control dandelions, especially those that contain 2,4-D. Fall is the best time to control perennial broadleaves like dandelion, and both amine or ester formulations of 2,4-D provide optimum dandelion control in the fall.

If dandelions are problematic in the spring, dandelion control is optimized in the cooler months (April) by using ester formulations of broadleaf herbicides, or by using florasulam (Defendor) (see below). However, even ester formulations can be ineffective if applied too early. As spring progresses and temperatures warm, turf managers should switch from ester formulations of broadleaf herbicides to amine formulations because (1) their efficacy improves during warmer temperatures in May and early summer and (2) they are safer than ester formulations to use around landscape plantings, gardens, and other sensitive plants. For more information about using amines or esters, see *Should I use an amine or ester formulation for postemergence broadleaf weed control?*, page 63.

While spring applications provide short-term control, dandelions often recover by mid-summer by regrowing from their taproot. As such, fall should always be the primary herbicide application timing for dandelion control.

For the lawn care industry, dandelions pose a problem in that they often flower in the spring before a broadleaf herbicide is applied. This reduces customer satisfaction because the customers see flowering dandelions (a weed they are paying professionals to remove). A typical scenario is that dandelions flower after a preemergence herbicide for crabgrass has been applied in the first round of treatments but before a broadleaf herbicide can be applied in the second round of treatments.

An early-season control option is florasulam. Florasulam (Defendor) is more effective at controlling dandelions in cool weather than traditional herbicides, and it provides the added benefit of preventing flowering in dandelion. Applying florasulam at 4 fl oz/A (0.013 lb ai/A) with a preemergence crabgrass herbicide at the typical late winter/early spring application timing, at least two weeks prior to dandelion flowering, will help control dandelions (although a broadleaf herbicide is still recommended later in spring for complete dandelion control). This application also should prevent dandelion from flowering, which should help keep homeowners happy, retain clients, and increase flexibility in making spring herbicide applications. Florasulam is also an effective herbicide when applied in the fall.

To learn which herbicides control dandelion best, see *Weed Control Ratings for Postemergence Broadleaf Herbicides*, page 76.

False-Green Kyllinga

Kyllinga species (*Kyllinga* spp.) are smaller sedge species that have compact seedheads. False-green kyllinga (*Kyllinga gracillima*)—see photos on page 14—is the most problematic of the Kyllinga species in the Midwest and Northeast. It is particularly common in the mid-Atlantic. Like yellow and purple nutsedges, kyllinga can be identified by its triangular stems and three-ranked leaf arrangement. False-green kyllinga spreads by rhizomes and seeds and is adapted to lower mowing heights than yellow nutsedge. It is often found in golf course fairways, tees, and putting greens.

In cool-season turfgrass, imazosulfuron (Celero) controls false-green kyllinga best. Sequential applications of halosulfuron (Sedgehammer) and pyrimisulfan (Arkon, Vexis) at maximum labeled rates provide suppression. Removal with a shovel or sod cutter is also effective because rhizomes are shallow enough for complete removal. The combination of an herbicide application followed by interseeding a desirable turfgrass at least one month after the herbicide is applied will result in the best control in heavily infested areas.

In warm-season turf, imazaquin (Image 70DG), imazosulfuron (Celero), sulfosulfron (Certainty), and trifloxysufluron (Monument, Recognition) are good options for false-green kyllinga control.

Fescue—*see Tall Fescue*

Field Paspalum—*see Dallisgrass*

Fountain Grass

Fountain grass (*Pennisetum* spp.—see photos on page 26) is an ornamental grass common to the landscapes of homes, commercial sites, and golf courses. However, following years of planting in these landscapes, we now realize that this species produces many viable seeds that drop onto the adjacent lawn and then become tough-to-control perennial grassy weeds. Although most of the ornamental grasses cannot withstand short mowing, fountain grass does. Typically, this grass can be identified by its visible shredded leaf blades in mid- and late summer, which leaves a whitish, wispy-looking clump when mown.

This weed can be controlled by applying herbicides that contain quinclorac (Drive XLR8 and others) and fenoxaprop (Acclaim Extra and Last Call). Two applications are needed for control. Glyphosate (Roundup and others) will also work as a nonselective spot-treatment option.

Goosegrass

Goosegrass (*Eleusine indica*—see photos on page 24) is best controlled with preemergence herbicides. Herbicides that contain oxadiazon (Goosegrass/Crabgrass Control, Ronstar, Oxadiazon) are the most consistent products for preemergence control of goosegrass. Other preemergence herbicides typically provide less goosegrass control. Anderson's Crab and Goose Control is labeled for putting green application.

Apply oxadiazon at the same preemergence application timing as a normal crabgrass application (the timing varies by location, but it is generally before April 15 in the Midwest and Northeast). In areas with severe goosegrass pressure, consider a split application approach similar to that described in the crabgrass control section. Make the initial application at the normal crabgrass preemergence timing at half the rate, and then apply the second (sequential or split) application 45–60 days after the first. Oxadiazon is less effective against crabgrass than other preemergence herbicides such as dithiopyr, prodiamine, and pendimethalin. If you have a history of crabgrass and goosegrass, use oxadiazon as well as one of these herbicides to control both weeds. Even though goosegrass germinates after crabgrass, apply preemergence herbicides before crabgrass and goosegrass germinate.

Note: Only granular formulations of oxadiazon are labeled for use in cool-season turf.

For postemergence goosegrass control in cool-season turf, use fenoxaprop (Acclaim Extra), fluazifop (Fusilade II), topramezone (Pylex), or MSMA (golf courses and sod farms only). More than one application of these herbicides may be required for postemergence control of tiller goosegrass. Of these, Pylex is the most effective on tillered goosegrass. Sulfentrazone (Dismiss) is also effective on pre-tillered goosegrass. These herbicides are labeled for specific turfgrass species or for use in specific areas. Consult the labels and identify your turfgrass species before application.

Among the cool-season turfgrasses, controlling goosegrass in creeping bentgrass tees and fairways is difficult because the herbicides that work best to control goosegrass often injure creeping bentgrass. Fenoxaprop (Acclaim Extra) is labeled for use in creeping bentgrass at the low label rate of 3.5 fl oz/A. At this rate, control of small (two-leaf or smaller) goosegrass plants can be achieved. Because goosegrass control at the 3.5 fl oz/A is only achieved on small plants, apply Acclaim Extra at two-week interval in summer to keep goosegrass out of problematic bentgrass areas.

Newly published research shows that 2,4-D + dicamba + MCPP + carfentrazone (SpeedZone) applied at 4–5 pt/A can also control goosegrass. While this is a broadleaf herbicide, research confirmed its ability to control goosegrass. A second application may be made 30 days after the initial application if necessary. Tank-mixing SpeedZone with Topramezone (Pylex) at 0.5 to 1.0

oz/A reduces goosegrass bleaching that occurs with Pylex alone and improves control.

Topramezone (Pylex) is labeled for use in creeping bentgrass at the reduced rate of 0.25 fl oz/A. This rate is extremely low and it will be necessary to measure the application precisely to avoid injuring creeping bentgrass. In addition, use caution when applying to avoid overlapping. Make repeat applications on a 21-day interval if needed.

While the above information is for tees and fairways, golf course superintendents in the transition zone often struggle to keep goosegrass out of their putting surfaces. Currently, no herbicides are labeled for selective postemergence control of goosegrass on creeping bentgrass putting greens. However, research trials often indicate that Acclaim Extra causes no appreciable creeping bentgrass injury when applied at 3.5 fl oz/A on putting greens.

SpeedZone can also control goosegrass without injuring creeping bentgrass. Do not attempt to use Pylex or SpeedZone on putting surfaces. With all herbicides, more injury is likely with spot treatments because it is challenging to calibrate the equipment and apply accurately.

More herbicides are registered for use on fairways and tees than on putting greens. Herbicide labels that neither restrict nor allow putting green use place all liability on the end user, so use caution with these products. For these reasons, only use products that have label instructions specifically for putting greens.

In warm-season turfgrasses, different herbicides are used for goosegrass control. Postemergence control in bermudagrass and zoysiagrass include foramsulfuron (Revolver) and Tribute TOTAL (thiencarbazone + foramsulfuron + halosulfuron). You can use Fusilade II and Acclaim Extra on zoysiagrass. These products work best when tank-mixed with 1 qt/A of triclopyr (Turflon Ester Ultra or Triclopyr 4). When controlling goosegrass, avoid applications to drought- or heat-stressed turf.

Ground Ivy

Ground ivy (*Glechoma hederacea*—see photos on page 17)—sometimes called creeping Charlie—is a tough-to-control broadleaf weed in the mint family usually found growing in the shade.

In cool-season turfgrass, herbicides that work best to control ground ivy include fluroxypyr and triclopyr (Turflon Ester Ultra, Triclopyr 4). Although there are many products that contain triclopyr in our resarch, triclopyr by itself at 1 qt/A typically provides the best results. Fluroxypyr (Vista XRT) by itself, or in combinations with other herbicides such as 2,4-D or triclopyr, is generally effective at controlling ground ivy.

Fluroxypyr and triclopyr work well as standalone herbicides because the rate of ingredient is higher in the standalone products than when they are part of a two-, three-, or four-way product. If you do not use Triclopyr 4, Turflon Ester Ultra, or Vista XRT then the following products should be good options because they all contain fluroxypyr, triclopyr, or both: 4-Runner, 4-Speed XT, Battleship III, Chaser, Chaser 2 amine, Confront, Cool Power, Escalade 2, Horsepower, Eliminate, Foundation, Momentum 4-Score, Momentum FX2, Sure Power, Tailspin, Turflon II Amine, and TZONE. Additionally, good ground ivy control can be achieved by tank-mixing Turflon Ester Ultra (or Triclopyr 4) at 1 pt/A tank-mixed with 3 pt/A of 2,4-D, or by tank-mixing Turflon Ester Ultra or Triclopyr 4 at 1 pt/A with your favorite three-way herbicide (2,4-D + MCPP + dicamba). Alternatively, you can acheive good ground ivy control tank-mixing Vista XRT at 11 fl oz/A with 3 pt/A of 2,4-D; or by tank-mixing Vista XRT at 11 fl oz/A with your favorite three-way herbicide (2,4-D + MCPP + dicamba).

Quinclorac (Drive XLR8, Eject 75DF, Quinclorac 75DF, Quinclorac 1.5L, QuinPro Herbicide) has been shown to be effective on ground ivy populations at most locations in most states, but it has not provided adequate control in Purdue University testing in West Lafayette, Indiana. Ground ivy populations are known to vary in their tolerance to herbicides, so it is wise to consider alternating herbicides or using tank-mixes when treating ground ivy.

In warm-season turf, metsulfuron (Manor, Mansion, MSM) or Celsius WG (thiencarbazone + iodosulfuron + dicamba) are other good options for ground ivy control.

Japanese Stiltgrass

Japanese stiltgrass (*Microstegium vimineum*) is a summer annual grassy weed commonly invasive to forests. Often, it is problematic in shaded lawn areas adjacent to an infested forest. Infestations in full-sun turfgrass areas, where you would normally expect crabgrass to be the predominant summer annual weed, are becoming more common. The life cycle and preemergence control options for Japanese stiltgrass are similar to those for crabgrass, except stiltgrass usually germinates a few weeks earlier than crabgrass. This complicates control with preemergence herbicides. While forsythia full bloom can be used as an indicator to apply crabgrass preemergence herbicides, there is no information available regarding soil temperatures or phenological indicators for Japanese stiltgrass emergence.

Do not expect seedlings to be killed by frost. Preemergence herbicides used to control crabgrass (dithiopyr, prodiamine, and pendimethalin) are effective for Japanese stiltgrass control if applied before stiltgrass germination.

A mixture of the postemergence herbicide fenoxaprop (Acclaim Extra at 9 fl oz/A) and a preemergence herbicide is effective for stiltgrass control when applied around forsythia full bloom. For stiltgrass control in summertime, higher rates of fenoxaprop (Acclaim Extra or Last Call) will be necessary for control. Fenoxaprop and topramezone (Pylex) are the only herbicides registered for selective postemergence stiltgrass control in cool-season turf. Of the two, fenoxaprop may work best. In fine fescues, sethoxydim (Segment II) may also be used.

Knotweed—*see Prostrate Knotweed*

Kyllinga—*see False-Green Kyllinga*

Lespedeza, Common or Annual

Lespedeza (*Lespedeza striata*—see photos on page 15) is often an indicator of insufficient nitrogen. Products that contain metsulfuron, fluroxypyr, or triclopyr are very effective on most legumes like lespedeza. 2,4-D alone will not control lespedeza or white clover.

Multiple products and product combinations that contain metsulfuron, fluroxypyr, or triclopyr are available. Metsulfuron (Manor, Mansion, MSM) is safe for use on Kentucky bluegrass (not mixtures of Kentucky bluegrass with perennial ryegrass or other cool-season turf species) when applied at 0.25–0.5 oz/A. When using three-way or four-way herbicides containing 2,4-D, MCPP, and/or dicamba, repeat applications are usually needed for complete control. Although clopyralid (Lontrel) provides excellent white clover control, it does not control lespedeza.

Moss

Silvery thread moss (*Bryum argenteum*—see photos on page 11) is the most common moss species found on golf course putting greens. Mosses are able to photosynthesize and fix nitrogen so are not parasitic. Moss mats tend to develop in wet, closely mown, heavily shaded areas. To reduce moss, consider increasing the light (by pruning trees), increasing air circulation (by pruning trees or installing fans), and raising the mowing height of putting greens. Moss can also occur on lawns and is limited to wet, shaded areas.

Moss control requires a persistent, long-term approach that combines cultural and chemical methods. It is important to realize that only a thick, healthy turf will cure and prevent recurrence of moss. The key objectives of a long-term approach are to reduce surface moisture and to improve both air circulation and light exposure. This may require removing adjacent underbrush and selective trees. Improve surface and subsurface drainage, decrease the frequency and amount of irrigation, and reduce the frequency or rate of nitrogen application.

Moss is enhanced by light and frequent foliar fertilization, especially when using ammonium sulfate as the nitrogen source, so consider fertilizing with granular nitrogen sources to help reduce moss populations.

If the moss covers a large area, spiking, verticutting, and topdressing will help break up and dry the moss mat. Moss that turns orange or golden brown indicates that desiccation is occurring. Raising the mowing height encourages turf to grow, which helps reduce moss.

Carfentrazone (Quicksilver T&O) applied at 2.0–6.7 oz/A in a spray volume of 100 gal of water/A when temperatures are less than 85°F on sunny days provides excellent moss control. However, annual bluegrass may be injured at rates greater than 2.0 oz/A. Creeping bentgrass has excellent carfentrazone tolerance.

To optimize moss control with carfentrazone, implement a program that includes frequent, light sand topdressing.

For applications to putting greens, there are isolated reports of some injury to creeping bentgrass and annual bluegrass if carfentrazone is applied within 60 days of an application of bensulide (Bensumec, Pre-San, Weedgrass Preventer, Goosegrass/Crabgrass Control). To reduce the risk of this occurring, avoid applying these herbicides within 60 days of one another.

Musk Thistle—*see Thistles*

Narrowleaf Plantain—*see Buckhorn Plantain*

Nimblewill

Nimblewill (*Muhlenbergia schreberi*—see photos on page 25) is a warm-season grass that turns brown at the first frost and greens up in the spring. It forms patches of thin turf in lawns that are most visible in early spring. Nimblewill spreads from seeds and stolons.

Nonselective and selective control options exist for this weed. Usually, turf managers attempt to control spreading grasses with a nonselective systemic herbicide like glyphosate (Roundup and others). Best results are seen when the weedy plants are young, fully green, and actively growing. At least two glyphosate applications are recommended, but three or more may be needed. You must allow the weed to regrow before making a follow-up application.

Mesotrione (Tenacity) can be used for selective control of nimblewill growing in a cool-season turf. To control nimblewill, start applying mesotrione in the spring (late April) with a nonionic surfactant at 0.25% (v/v). You will need to make two or three applications. If you make three applications, use the 5 fl oz/A rate (or 6, 6, and 4 fl oz/A). The product does have a yearly maximum use rate, and the label states "do not apply more than 16 oz of Tenacity per acre per year or per crop (equivalent to a maximum of 0.50 lb of mesotrione per acre per year)." You can also apply mesotrione in late summer and fall, but you should initiate them by August for best results.

Another option is to apply topramezone (Pylex) at 1–1.5 fl oz/A at 21- to 28-day intervals starting in late April. For best results, include a methylated seed oil at 0.5%–1.0% (v/v).

Orchardgrass

Orchardgrass (*Dactylis glomerata*—see photos on page 22) is a bunch-type, cool-season grass often found in rural lawns in the Midwest. Orchardgrass is commonly used in pastures for grazing, but it doesn't make a particularly dense lawn. It is light blue-green and is spotted in lawns when it grows more quickly than the surrounding turf.

Unfortunately, only nonselective control options exist for this weed in cool-season turf. We have tested other herbicides such as mesotrione (Tenacity), chlorsulfuron (Telar XP), metsulfuron (Manor, Mansion, MSM), and thiencarbazone + iodosulfuron + dicamba (Celsius) on orchardgrass, but they did not provide satisfactory control. Orchardgrass can be controlled with a nonselective systemic herbicide like glyphosate (Roundup and others). Another option is to use a knife to physically/mechanically remove the orchardgrass.

Oxalis—*see Yellow Woodsorrel*

Poa annua—*see Annual Bluegrass*

Poa trivialis—*see Rough Bluegrass*

Prostrate Knotweed

Prostrate knotweed (*Polygonum aviculare*—see photos on page 29) is an early germinating summer annual broadleaf that is often found in low-oxygen soils, including compacted areas next to sidewalks and previously flooded areas. Germination starts in late February and early March in many Midwest and Northeast states.

By itself, 2,4-D will provide only fair control of prostrate knotweed, but when combined with triclopyr (Turflon Ester Ultra or Triclopyr 4) or dicamba (Banvel, Vanquish), it should provide excellent control in cool-season turf. There are many combination products that contain 2,4-D and dicamba, including Trimec 992 and SpeedZone. Combination products that contain 2,4-D and triclopyr include 4-Speed XT, Chaser, Chaser 2 Amine, Momentum FX2, Sure Power, Turflon II Amine, and TZONE.

In warm-season turf, metsulfuron (Manor, Mansion, MSM), metsulfuron + dicamba (Fahrenheit), or the herbicides listed above will provide effective postemergence control of prostrate knotweed.

Preemergence control of prostrate knotweed can be achieved with late fall (November or December) applications of isoxaben (Gallery, Isoxaben 75WG) or isoxaben-containing herbicides such as Crew (dithiopyr + isoxaben) or Gemini (prodiamine + isoxaben). Other preemergence herbicides will also work but are less effective than isoxaben. You can apply isoxaben in late winter, but spraying conditions are not typically favorable at that time of year and it is difficult to predict exactly when prostrate knotweed might germinate, although it is usually in early March.

Prostrate Spurge

Prostrate spurge (*Euphorbia maculata*—see photos on page 14) is a summer annual that germinates in June and July and can be controlled with three- or four-way herbicide mixtures in cool-season turf. Repeat applications are usually needed both from lack of complete control and germination of new seedlings.

You may improve control by using herbicides containing sulfentrazone, carfentrazone, dichlorprop (2,4-DP), and triclopyr. Preemergence control with dinitroaniline herbicides (Barricade, Pendulum) is inconsistent, but it is more consistent with isoxaben (Gallery) or isoxaben-containing herbicides such as Crew (dithiopyr + isoxaben) or Gemini (prodiamine + isoxaben). In warm-season turf, metsulfuron (Manor, Mansion, MSM) or Blindside (sulfentrazone + metsulfuron) are the best postemergence control treatments.

Purple Nutsedge

Purple nutsedge (*Cyperus rotundus*—see photos on page 14) is less common than yellow nutsedge. Because purple nutsedge is not as cold hardy as yellow nutsedge, it typically occurs in areas where bermudagrass and zoysiagrass are grown.

Purple nutsedge has darker leaf blades than yellow nutsedge, and it has a more blunt leaf tip. Purple nutsedge has a purple to maroon seedhead—hence, its name. It is controlled with similar herbicides used to control yellow nutsedge, but purple nutsedge is more difficult to control with these products.

In warm-season turfgrasses, sulfonylurea herbicides including Celero (imazosulfuron), Certainty (sulfosulfuron), and Monument or Recognition (trifloxysulfuron), are effective. Dismiss South (sulfentrazone + imazethapyr) is also effective at controlling purple nutsedge in warm-season turf.

Purslane

Purslane (*Portulaca oleracea*—see photos on page 31) is a summer annual broadleaf weed that is problematic during spring and summer turfgrass establishment in bare soil areas and other areas next to sidewalks and driveways. Germination starts in May in many Midwest states and continues into August.

Triclopyr (Turflon Ester Ultra or Triclopyr 4), fluroxypyr (Vista XRT), and dicamba (Banvel, Vanquish) are the best herbicides for postemergence purslane control. There are many combination products that contain one or more of these ingredients. For a list of products containing these ingredients, see *Commonly Used Broadleaf Herbicide Combinations for Turfgrass*, page 31.

In warm-season turf, metsulfuron (Manor, Mansion, MSM), Fahrenheit (dicamba + metsulfuron), or the herbicides listed above will provide effective postemergence control of purslane.

Preemergence control of purslane can be achieved with spring applications of isoxaben (Gallery, Isoxaben 75WG) or isoxaben-containing herbicides such as Crew (dithiopyr + isoxaben) or Gemini (prodiamine + isoxaben) in cool-season turf, or with late-winter applications of simazine (Princep Liquid and others) in warm-season turf. Other preemergence herbicides will also work but are less effective than isoxaben.

Quackgrass

Quackgrass (*Elymus repens*—see photos on page 24) is a rhizomatous cool-season grass that forms thin patches in desirable turf species. It has an extensive rhizomatous growth habit that invades both turf and landscape beds, and it is easily distinguished by its long-clasping auricles.

Turf managers traditionally attempt to control spreading grasses like quackgrass with a nonselective systemic herbicide like glyphosate (Roundup and others). Results are best when applications are made while the weedy plants are young, fully green, actively growing, and not under drought stress. At least two glyphosate applications are recommended, but three or more may be needed. You must allow the weed to regrow before making a follow-up application. There are currently no selective herbicides labeled for removing quackgrass from cool-season turf, except fine fescues.

Rough Bluegrass

Rough bluegrass (*Poa trivialis*—see photos on page 27), sometimes called roughstalk bluegrass, is a perennial grass that spreads by stolons. This grass is typically considered a weed because it grows well in the spring and fall but is very sensitive to heat, drought, and disease in the summer months and will die or lie dormant in the summer until regrowth occurs in September. This weed is problematic in creeping bentgrass fairways and Kentucky bluegrass lawns and athletic fields and often becomes established in these areas initially as a contaminant in seed lots. Raising the mowing height (3 inches tall) will help to reduce rough bluegrass over time.

Bispyribac-sodium (Velocity PM) is registered for rough bluegrass control in athletic fields, lawns, golf courses, and sod farms. Application timing for Velocity PM is between late spring and early fall when daytime highs are between 70°F and 85°F. For controlling rough bluegrass in creeping bentgrass fairways, follow the same guidance as for *Poa annua* control on page 27. For rough bluegrass control in perennial ryegrass or tall fescue, follow guidance on page 42. Velocity PM is capable of causing severe injury to certain cultivars of Kentucky bluegrass (*Poa pratensis*). As always, read the complete label instructions before making an application.

To control rough bluegrass in cool-season lawns and athletic fields, nonselective control with glyphosate (Roundup and others) followed by reseeding is an option. Control with glyphosate is optimal when it is applied in the spring before summer stress. If the areas are small, consider mechanically removing the areas with a shovel or a sod cutter, taking care to remove the patch and at least 12 inches of turf surrounding the patch. Removing about 0.5 inch of soil should sufficiently remove this weedy grass because its growing points (crowns and stolons) are all at the soil surface. Reseed or sod the area with a desirable turfgrass species after removal.

To control rough bluegrass in warm-season lawns and athletic fields, apply a nonselective herbicide like glyphosate in late winter while bermudagrass is dormant. Alternatively, many sulfonylurea herbicides are labeled for the selective removal of cool-season grasses such as rough bluegrass from bermudagrass and zoysiagrass. See *How can I safely remove overseeded turf (that is, perennial ryegrass) without injuring the bermudagrass?*, page 65 for more information.

Sandbur

Longspine sandbur (*Cenchrus longispinus*—see photos on page 22) and field sandbur (*Cenchrus incertus*) are summer annual grasses often found in sandy soils, but they can grow in a range of soil textures. Sandburs appear similar to crabgrass (*Digitaria* spp.) and foxtails (*Setaria* spp.). Sandbur seedheads are spikelike racemes with bur-like fruit. These burs can cause physical injury or discomfort when people step or fall on them in parks or athletic fields.

To control sandbur with preemergence herbicides, follow the same recommendations given for crabgrass control. Among the postemergence herbicides, those containing fenoxaprop (Acclaim Extra, Last Call), MSMA, or topramezone (Pylex) are most effective.

Spurge—*see Prostrate Spurge*

Tall Fescue

Tall fescue (*Schedonorus arundinaceus*—see photos on page 28) is a coarse-textured (wide-bladed), bunch-type grass that is often used as a primary turf species. Improvements in tall fescue leaf texture have resulted in more fine-bladed varieties/cultivars that have helped to increase the turf quality and use of this drought-tolerant turf.

We often recommend improved turf-type varieties of tall fescue as the primary species for lawns, but older, wider-bladed, forage-type varieties of tall fescue such as Kentucky 31 (KY-31) still exist and can be problematic weeds in lawns. This forage-type tall fescue can be easily spotted in lawns in early spring when they greens up faster than Kentucky bluegrass and by its bunch-type growth habit, which results in a clump of turf when present as a weed.

When there are few weedy patches of bunch-type grasses like tall fescue, it is best to cut them out with a shovel or knife. Be sure to cut down a couple of inches into the soil to get all the stems. Refill the holes with soil and seed immediately with the desirable turf species similar to the existing turf.

If the area has a large number of tall fescue plants, chemical control will be more efficient. A nonselective herbicide such as glyphosate (Roundup and others) can be spot-applied. These herbicides will also kill the surrounding desired turf species, so be careful to treat only the clump of tall fescue. Two applications may be needed for complete control, and the area should be reseeded shortly after the final application.

Chlorsulfuron (Telar XP and Chlorsulfuron 75DF) is a selective herbicide registered for tall fescue control. Chlorsulfuron was previously sold as a product called Corsair, which was a selective herbicide registered for tall fescue control in many turf species. Some quantities of Corsair (labeled for most major turf areas) may still exist, but this herbicide is no longer sold. Telar XP and Chlorsulfuron 75DF contain the same ingredient as Corsair, but applicators can only use them on turf in industrial sites and roadsides, not in managed turf areas like golf courses, lawns, athletic fields, and sod farms. Spot-apply chlorsulfuron according to label instructions. The tall fescue will slowly thin and die, allowing the desired species to fill in. Reseeding should not be needed if Kentucky bluegrass is present. Chlorsulfuron will kill perennial ryegrass, so do not apply it to perennial ryegrass.

To control tall fescue in warm-season lawns and athletic fields, apply a nonselective herbicide like glyphosate in late winter while bermudagrass is dormant. Alternatively, many sulfonylurea herbicides are labeled for the selective removal of cool-season grasses such as tall fescue from bermudagrass and zoysiagrass. For more information, see *How can I safely remove overseeded turf (that is, perennial ryegrass) without injuring the bermudagrass?*, page 65.

Thistles

Canada thistle (*Cirsium arvense*—see photos on page 8), musk thistle (*Carduus nutans*—photos on page 7), and bull thistle (*Cirsium vulgare*—photos on page 8) are all common turf weeds in the Midwest. Of these, Canada thistle is typically the most problematic. Bull and musk thistles are biennials, whereas Canada thistle is a rhizomatous perennial.

Herbicides that contain clopyralid (such as 2-D, Confront, Lontrel, Millennium Ultra 2) in nonresidential turf, or mixtures that contain 2,4-D, dicamba, or triclopyr in residential turf, typically provide the best thistle control. Regardless of herbicide selection, thistle may require multiple herbicide applications for complete control. NativeKlean (2,4-D + aminopyralid) is also an effective option for thistle control in native or natural areas that are not regularly mowed or maintained.

Violet—*see Wild Violets*

White Clover

White clover (*Trifolium repens*—see photos on page 16) is one of the most common weeds in cool-season turfgrasses. It tolerates close mowing, grows in various soil types, and can fix its own nitrogen. White clover spreads by stolons and seed.

While most view white clover as a weed, some consider it an important component of the turf that adds beneficial soil nitrogen. Early lawn seed mixtures in the Unites States included white clover, and today, some grass-legume seed mixtures are being sold again. Still, many want to keep white clover out of their turf.

Some herbicides (such as 2,4-D, glyphosate, and sulfentrazone) will not effectively control white clover. In addition, applying herbicides in spring or early summer provides poor control compared to fall applications. Herbicides containing clopyralid, dicamba, fluroxypyr, florasulam, and/or quinclorac provide the best white clover control, especially when applied in fall.

Wild Garlic

Wild garlic (*Allium vineale*—see photos on page 19) and wild onion (*Allium canadense*—see photos on page 18) are similar weed species in turf. These bulb-forming weeds are most obvious in the winter when turf is not growing and not being mown. Wild garlic can be distinguished from wild onion by its hollow stem—wild onion has a solid stem.

In cool-season turf, wild garlic is more difficult to control. Use 2,4-D or one of the various combinations of 2,4-D, MCPP, and dicamba (Trimec, Triplet, others) for fair control. These products provide limited wild garlic control. Treatments are optimal in late fall and early spring when wild garlic foliage is most prevalent. The ester formulations of 2,4-D are more effective against garlic than amine formulations (see *Should I use an amine or ester formulation for postemergence broadleaf weed control?*, page 63). Additionally, mowing prior to application may improve herbicide uptake and control.

In warm-season turf, metsulfuron (Manor, Mansion, MSM), Blindside (sulfentrazone + metsulfuron), and sulfosulfuron (Certainty) are very effective for wild garlic control. Apply in late March on a warm (at least 50°F), sunny day when there is good soil moisture. Metsulfuron also controls most of the common winter broadleaf weeds such as henbit and common chickweed. Add 0.25% (v/v) nonionic surfactant to metsulfuron and sulfosulfuron. Unlike some phenoxy herbicides, metsulfuron does not delay spring transition of partially green bermudagrass.

Wild Onion—*see Wild Garlic*

Wild Violets

Collectively, turf managers refer to the Midwest species common blue violet (*Viola sororia*), wooly blue violet (see photos on page 35), and confederate violet (*Viola sororia priceana*—photos on page 35) all as wild violet. Wild violets are a persistent, perennial, and difficult-to-control broadleaf weed.

Good control of wild violet is typically obtained with triclopyr (Turflon Ester Ultra or Triclopyr 4). Many herbicides are available with triclopyr as the key ingredient (see *Commonly Used Broadleaf Herbicide Combinations for Turfgrass*, page 95). For best control with triclopyr use more than 0.5 lb ae/A. Turflon Ester Ultra and Triclopyr 4 at 1 pt/A will deliver 0.5 lb ae/A; and at 1 qt/A will deliver 1.0 lb ae/A. Chaser, Chaser 2 Amine, Confront, 2-D, and Tailspin will also deliver ≥0.5 lb ae/A. To see the rates of triclopyr delivered by various products at the low and high label rates, see *Active Ingredients in Commonly Used Postemergence Herbicide Combinations*, page 97.

In bermudagrass and zoysiagrass, metsulfuron (Manor, Mansion, MSM), sulfentrazone + metsulfuron (Blindside), and flazasulfuron (Katana) are effective violet control herbicides. Repeat applications are often required regardless of the herbicide. Fall applications are best followed by spring applications.

Windmillgrass

Windmillgrass (*Chloris verticillata*— see photos on page 22) is also called tumble windmillgrass. It gets its name from its globe-shaped panicle that easily tumbles from one location to another with the wind and spreads seed.

Windmillgrass is found in poor-quality soils such as roadsides, parking lot islands, and other areas with droughty soils. Topramezone (Pylex) + triclopyr provides the best control of this perennial grass weed. Mesotrione (Tenacity) and Acclaim Extra (fenoxaprop) are additional options for windmillgrass control. More than one application of these herbicides may be required for postemergence control of windmillgrass. Make sure to include the label-specified adjuvant in the spray tank when applying these herbicides. Adding triclopyr (Turflon Ester Ultra or Triclopyr 4) at 1 qt/A to either topramezone, fenoxaprop, or mesotrione will significantly improve windmillgrass control. This grass is quick to green up in spring but enters winter dormancy early, so for best results, target applications in the late spring and early summer, similar to nimblewill control.

Woodsorrel—*see Yellow Woodsorrel*

Yellow Nutsedge

Many herbicides are available for yellow nutsedge (*Cyperus esculentus*—see photos on page 13) control, but herbicide application timing is critical to optimize control. Herbicide applications made before the initiation of tuber production in midsummer will help reduce the severity of this weed. However, the most common mistake is to make herbicide applications too late in the season after yellow nutsedge is big and is spreading by rhizomes and producing tubers. Prevention is the key cultural control for yellow nutsedge. A good yellow nutsedge herbicide control program will need to be implemented early in the season and in consecutive years.

The options for yellow nutsedge control are different for cool-season and warm-season turf areas.

In cool-season turf, products that contain sulfentrazone (Dismiss) can also provide preemergence and postemergence control of annual sedge and yellow nutsedge, although only Echelon (prodiamine + sulfentrazone) is labeled for preemergence control. Dismiss is the primary postemergence herbicide with sulfentrazone, although Solitare (sulfentrazone + quinclorac) has similar amounts of sulfentrazone. Q4 Plus, Surge, SureZone, and TZONE are among the many herbicides that also contain sulfentrazone but are labeled for yellow nutsedge suppression, not control, because these products contain a lower amount of sulfentrazone in their formulations (see page 104). Of the various sedge control herbicides, sulfentrazone will provide the quickest control of sedges—injury will appear within a few days of application. The rate of sulfentrazone will affect the level of control but not the speed of activity. When using products that contain sulfentrazone, do not add a surfactant.

Halosulfuron (SedgeHammer, ProSedge, and others) provides excellent yellow nutsedge control with very good turfgrass tolerance. When using products containing 75% halosulfuron (that is, 75DF formulation), add a nonionic surfactant to improve control. Injury to yellow nutsedge will appear about two weeks following application. For spot treatment, mix 0.9 gram of halosulfuron in 1 gallon of water with 0.33 fl oz of nonionic surfactant or 0.25% (volume of surfactant per volume of water [v/v]) nonionic surfactant. For larger areas use 0.66 to 1.33 oz/A. A SedgeHammer+ formulation also is available that already includes surfactant, so adding surfactant to this product is not necessary. For spot treatment with SedgeHammer+, use 0.5 oz/1,000 ft^2.

Imazosulfuron (Celero) also provides excellent yellow nutsedge control with very good turfgrass tolerance and is labeled for all major turfgrass species except annual bluegrass. Apply 8 to 14 oz of Celero with a nonionic surfactant at 0.25% (v/v) for the control of sedges and kyllinga.

Mesotrione (Tenacity) is labeled for postemergence control of yellow nutsedge in Kentucky bluegrass, perennial ryegrass, tall fescue, and fine fescue. This herbicide causes a bleaching effect on susceptible weeds. Adding a nonionic surfactant according to label recommendations will improve control with mesotrione. Two applications (applied two weeks apart) will be required for control.

In warm-season turf areas (such as bermudagrass and zoysiagrass), any of the cool-season sedge herbicides mentioned above will control yellow nutsedge (check the label to make sure they can be used on your specific turf species). In addition, other sulfonylurea herbicides including Monument or Recognition (trifloxysulfuron), Katana (flazasulfuron), or Certainty (sulfosulfuron) may be used in bermudagrass and zoysiagrass. Injury to yellow nutsedge will appear about two weeks after application. Repeat applications of these herbicides will often be needed if regrowth appears. Make sure to add 0.25% (v/v) nonionic surfactant to these sulfonylurea herbicides.

Other products—including Basagran T/O (bentazon), Blindside (sulfentrazone + metsulfuron), Tower (dimethenamid), FreeHand (dimethenamid + pendimethalin), Pennant MAGNUM (metolachlor), and Tribute TOTAL (thiencarbazone + foramsulfuron + halosulfuron) —are labeled for yellow nutsedge control in bermudagrass and zoysiagrass.

Yellow Woodsorrel

Yellow woodsorrel (*Oxalis stricta*—see photos on page 20) and creeping woodsorrel (*Oxalis corniculata*) are similar species that can be found in turf. Yellow woodsorrel is more common in turf; creeping woodsorrel is more common in landscape plantings.

Typically, the most effective postemergence products contain fluroxypyr (Battleship III, Escalade 2, Momentum 4-Score, Momentum FX2, Tailspin, Vista XRT), triclopyr (2-D, 4-Runner, 4-Speed XT, Battleship III, Chaser, Chaser 2 Amine, Confront, Cool Power, Eliminate, Horsepower, Momentum 4-Score, Momentum FX2, Tailspin, Three-Way Ester II, Turflon Ester Ultra, Turflon II Amine, Triclopyr 4, TZONE), or both fluroxypyr and triclopyr (Battleship III, Momentum 4-Score, Momentum FX2, Sure Power, Tailspin).

Zoysiagrass

Zoysiagrass (*Zoysia* spp.—see photos on page 29) is a stoloniferous and rhizomatous warm-season grass that forms dense patches of turf. Under the right circumstances, it is considered a desirable turf species, but when it invades a lawn as a patch or spreads into the periphery of a cool-season lawn it can be considered a weed.

Zoysiagrass has an extensive system of stolons above ground and rhizomes below ground that make it difficult to control. Traditionally, turf managers attempt to control spreading grasses such as zoysiagrass with a nonselective systemic herbicide like glyphosate (Roundup and others). Best results are seen when applications are made when the zoysiagrass is fully green, actively growing, and not under drought stress. At least three glyphosate applications are recommended. You must allow the weed to regrow before making a follow-up application. If you desire optimum control, this will leave the area dead and unsightly for months. Physically removing the soil (including the rhizomes and stolons) may be needed for complete control, but is very difficult.

There are currently no selective herbicides that provide excellent control of zoysiagrass in cool-season turf. However, similar to bermudagrass, applying topramezone (Pylex) at 1.33 fl oz/A at 21-day intervals staring in late summer will suppress zoysiagrass. Use topramezone in conjunction with an overseeding of a desirable cool-season turfgrass after the last application and repeat for two consecutive years—this strategy can selectively work to remove zoysiagrass. Expect 50%–75% zoysiagrass removal per year with this topramezone + overseeding strategy.

Frequently Asked Questions and Answers About Weed Control with Herbicides

Contents

General Information

Q: How can I get another copy of this publication?

A: This publication is updated annually to provide current recommendations to professional turfgrass managers. The most up-to-date version of this publication is available from the Purdue Extension Education Store (*https://.edustore.purdue.edu*, (765) 494-6794), and participating extension services (see *Where to Find Additional Resources*, page 123). For updates, corrections, and other notes, visit *https://turf.purdue.edu/extpub/turfgrass-weed-control-for-professionals*.

Updates to the website will be posted as needed.

Q: What new herbicides are covered in the 2024 edition?

A: New ingredients, formulations, or mixtures to the 2024 edition are added below. Not included in this table are new trade names (generics) of products included in previous editions.

Turf Herbicides Added in This Edition

Common Name	Trade Name
bispyribac-sodium	Velocity PM
flumioxazin + pyroxasulfone	SureGuard Xtra
MCPA + triclopyr + dicamba + sulfentrazone	4-Runner
pyrimisulfan	Arkon

Q: Are there any herbicides not included in this publication?

A: This publication only includes herbicides registered for use in turf (golf courses, commercial and residential lawns, sod farms, athletic fields, grounds, parks, industrial sites, cemeteries, and hospitals). It does not include herbicides labeled for use in other crops, or land and vegetation management.

Almost all turf herbicides used in the United States are registered in the North Central Region and are listed in this guide. A few herbicides are registered only within specific states in the Northeast and North Central regions. Check with your local distributors for availability. Below are some herbicides that are not registered for use in any of the states in our region but are used by turf managers in other regions of the United States. These products may be of interest to turf managers in the transition zone or in the southern or western United States.

Turf Herbicides Not Labeled for the Midwest or Northeast

Common Name	Trade Name
2,4-D	Dri-Clean[1]
diclofop	Illoxan 3EC[2]
dimethenamid + pendimethalin	FreeHand CA 1.75 G[3]
metsulfuron + pyraflufen	Caliente WDG[2]
sulfentrazone	Zeus[3,4]

1 California, Idaho, Oregon, and Washington only.
2 Labeled in most of the southern United States. Discontinued product.
3 California only.
4 Sod farms only.

Q: How can I get more information about current turf problems and issues?

A: Various turf programs maintain electronic newsletters (email, listservs, or blogs) and/or distribute timely, helpful information via X (formerly Twitter). Check out the websites of the turf programs involved in this publication (see *Turf Programs*, page 122). Visit their websites to subscribe to their newsletters and learn more about following them on social media.

Q: Can I use the same herbicides on both commercial and residential turf?

A: Some herbicides are labeled for use in specific turf areas, although most can be used in lawns (commercial and residential), sod farms, golf courses, cemeteries, rights of way, sports fields, and parks. The *Herbicides Registered for Use in Specific Turf Sites* table (below) lists herbicides that have labels that limit their use to very specific sites.

Herbicides and PGRs Registered for Use in Specific Turf Sites

Label Site Specifications	Herbicide/PGR
For sod farms only	Addax, Aim EC, Antik EC, Barrage HF, Clarity, Detonate, Diakon, Hi-Dep, Profine75, Rifle, Roundup WeatherMax, Sandea, Spartan 4F, Spartan Charge, Strut, Topeka, Yukon
For golf courses only	Goosegrass/Crabgrass Control, PoaCure
For golf courses and sod farms only	912 Herbicide, MSMA 6.6, MSMA 6 Plus, MSMA Herbicide, TARGET 6.6, TARGET 6 Plus
For use in unimproved industrial turf, on roadsides, and on other noncrop sites only	Chlorsulfuron 75DF, Derigo, E-2, NativeKlean, Patriot, Telar XP
Not for use in residential turf	All of the herbicides above, plus Broclean, Buctril, Buctril 4EC, Cardinal, Confront, Dolerity, Ethephon 2, Ethephon 2SL, GameOn, Kerb 50WP, Kerb SC T&O, Lontrel, Millennium Ultra2, Moxy 2E, Musketeer, Oskie, Oxadiazon 2G, Oxadiazon SC, Oxadiazon 50WSB, Rimsulfuron 25DF, Ronstar FLO, Ronstar WSP, Starfighter L, Trimmit 2SC, Trycera, Verve
Not for use in home lawns but can be used in residential landscapes	Lontrel

Q: Are herbicides toxic?

A: The toxicity of herbicides varies from one product to the next. Remember that herbicides are registered for use only after rigorous testing and approval by the US EPA. The signal word (danger, warning, or caution) on the herbicide provides information about the mammalian toxicity of these products. When used according to label instructions, herbicides are safe, effective, and pose little risk to humans, wildlife, or the environment.

Most problems caused by pesticides are due to improper use. More information about herbicide toxicity can be found in *Safety of Herbicides Compared to Other Commonly Used Chemicals* (University of Tennessee Extension publication, *chrome-extension://efaidnbmnnnibpcajpcglclefindmkaj/https://uthort.tennessee.edu/wp-content/uploads/sites/228/2023/11/W270-%20Herbicide%20safety%20comparison.pdf* .

Q: What "organic" herbicides or nonpesticide products are available for weed control?

A: Some organic or natural herbicides are available. An organic product is, by definition, a compound that contains carbon. However, more often than not, the term "organic" is used to describe a compound that is plant- or animal-based.

The predominant organic herbicide in turfgrass systems is corn gluten meal for preemergence crabgrass control. This byproduct of corn processing contains nitrogen and has shown to be effective at reducing crabgrass in lawns in northern states, although tests in Indiana and states south of Indiana show it has limited efficacy on crabgrass.

Among the postemergence organic herbicides, the most common are pelargonic acid (Scythe) and acetic acid (5% or greater solutions). Other products that contain medium-length fatty acids and clove oil (eugenol) show some promise; however, these organic postemergence herbicides are nonselective and can injure actively growing turfgrasses, so their use should be limited to directed spot treatments. The bottom line is that most organic postemergence herbicides have limited use in turf and are better suited to weed control in parking lots, fencerows, and other bare ground applications.

Many new natural weed control products contain the active ingredient iron HEDTA (FeHEDTA), including the professional product Fiesta. These products quickly injure broadleaf weeds and injure turf little, but they require multiple reapplications to provide effective control.

Q: How do I calculate how much a product costs per acre or per 1,000 square feet to make good purchasing decisions?

A: Information about this and other herbicide calculations is provided in *Pesticide and Plant Growth Regulator Math*, page 118.

Q: How do I convert from pounds (lb) active ingredient (ai) per acre (A) to the amount of product I need?

A: Information about this and other herbicide calculations isprovided in *Pesticide and Plant Growth Regulator Math*, page 118.

Weather Effects on Applications

Q: What is the best time of year to control weeds?

A: The optimum time to control weeds depends on the type and life cycle of the weed and, in some cases, the specific weed and specific herbicide—so this is not a simple question to answer. Regardless of the timing and weed, it is important to select herbicides with a high efficacy for the weeds you are targeting. Specific information about the optimal products and applications for various tough-to-control annual and perennial grasses is discussed in *Control of Tough Weeds*, page 42.

In the Midwest, perennial broadleaf weeds are best controlled with postemergence herbicides in the fall (September 1 to November 15). October applications are optimal. This window can be extended longer if the end of the fall is mild or if the turf is farther south. Herbicides are effective on perennial broadleaf weeds in fall because plants are more likely to translocate (move downward) herbicides into root and stem tissues as days get shorter and temperatures cool. Typically, this will occur near or following the first frosts.

Previous research in Nebraska demonstrates that 2,4-D and dicamba are far more effective at controlling dandelions and Canada thistle when applied 1 to 10 days after the first fall frost than when applied 5 to 11 days before the frost. Other work at Michigan State University found that "good dandelion control can result from herbicides applied through late October, even when the plants are not actively growing." More recently, Purdue researchers found that early to mid-November applications of most herbicides resulted in ground ivy control similar to earlier (September and October) applications.

Spring applications will also control perennial broadleaf weeds but are generally less effective than fall applications. See *Should I use an amine or ester formulation for postemergence broadleaf weed control?*, page 63.

Winter annual broadleaves (see *Weed Life Cycles*, page 3) can be controlled at the same fall timing as perennial broadleaves. Fall timings are best for winter annuals because they are small after just germinating and herbicides translocate well in the fall. Spring control of winter annual broadleaves is more difficult because the weeds are larger and are less likely to translocate herbicide. Additionally, controlling winter annuals in late spring is not typically necessary because these weeds will complete their life cycle and naturally die by early summer.

Summer annual broadleaves and summer annual grasses are best controlled with preemergence herbicides applied in the spring before they germinate. They can also be controlled with postemergence herbicides in early summer before they become too large and before they produce seed. For more specific information about preemergence control and postemergence control for problematic summer annual grasses like crabgrass, see *Control of Tough Weeds*, page 42.

Perennial grasses are difficult to control and herbicide application timing and selection are very specific to the target grass. For more specific information about controlling problematic perennial grasses like bermudagrass, creeping bentgrass, nimblewill, and more, see *Control of Tough Weeds*, page 42.

Many herbicides are available for nutsedge control, but proper herbicide use and application timing are critical to optimize control. Herbicide applications made before nutsedge plants produce tubers (typically in late summer) will help reduce the severity of this weed. The most common mistake is to apply herbicides too late in the season after yellow nutsedge is large and is spreading by rhizomes and producing tubers. A good nutsedge control program will need to be implemented early in the season and continue for consecutive years to reduce tuber populations in the soil and prevent the spread of this problematic weed.

Q: *What happens if it rains after I apply herbicides?*

A: Most postemergence herbicides should work effectively if not washed off the weed by rain or irrigation within about six hours after application. The actual amount of time varies from product to product. If it rains less than six hours after a postemergence application, control will likely be reduced depending on how soon it rained and the intensity of the rainfall. Preemergence herbicides actually require watering in within a few days after an application, so there is no concern if it rains following an application of a preemergence herbicide.

Q: *Can I apply herbicides when temperatures are 85°F or above?*

A: You will need to follow the herbicide label. Herbicide applications when daily high temperatures are above 85°F generally increase the risk of turf injury. Most herbicide labels do not prohibit applications at temperatures greater than 85°F or 90°F (each herbicide has a different maximum temperature), but they do caution against their use at these temperatures due to the increased risk for turf injury.

A general recommendation is to make the application if the site is more weeds than turf and you are willing to assume the risk of injuring the turf. In most cases, if there are few weeds in the turf, then it is best to delay making the application until cooler conditions exist.

Q: *What is a growing degree day (GDD)? How can I use this value to improve my weed control and my PGR applications?*

A: A growing degree day (GDD) is a method to track the heat units that have accumulated and are needed for plant growth and development. Models can be calculated in either Fahirenheit or Celsius. The formula for calculating GDD is

GDD = [(max temperature °F + min temperature °F) ÷ 2] - base temperature °F

The base temperature is the temperature where metabolism is minimal. It varies by model but is usually 22°F, 32°F (0°C), or 50°F (10°C). The GDD models for plant growth regulators are frequently in degrees Celsius with base temperatures of 0°C for cool-season grasses and 10°C for warm-season species.

For example, if the high today was 74°F and the low was 52°F and we use a base temperature of 50°F, our calculation would be

GDD = [(74°F + 52°F) ÷ 2] - 50°F = 13

On a cooler day, you might have a high of 58°F and a low of 40°F. Using the base temperature of 50°F, our calculation would be

GDD = [(58°F + 40°F) ÷ 2] - 50°F = -1

When a GDD is less than zero, we simply make it a zero and determine that no growth (or plant development) occurred on that day.

Models that help us predict plant development and growth use accumulated GDDs, which simply adds the GDDs calculated each day.

In some cases, accumulated GDDs can be used to monitor when weeds might germinate or flower or when grasses might produce a seedhead. In other cases, accumulated GDDs can be used to help optimize application timing such as with preemergence crabgrass applications or the selection of amines or esters for spring broadleaf applications, or when to apply a repeat application of a plant growth regulator. Research into plant development and optimal herbicide application timing has determined a window of accumulated GDD needed to best predict when to time these applications or when these events might occur.

While it is possible to track your own GDDs using a local weather station, an easier way to do this is to use an online tool. The Chicago District Golf Association, Illinois Turf Foundation, Michigan Turfgrass Foundation, Midwest Regional Turf Foundation, Ohio Turf Foundation, Project Green, and Purdue University Turf Science Program help support Michigan State University Extension's GDD Tracker, https://gddtracker.msu.edu/. This website tracks GDD for the Midwest and provides updates for golf courses on timing annual bluegrass seedhead suppression applications, updates for all in the turf industry on crabgrass germination and preemergence herbicide application timing, and updates on broadleaf flowering and the timing of amines and esters for spring broadleaf applications.

A web-based application, GreenKeeper, helps schedule follow-up applications of plant growth regulators. The tool (https://www.greenkeeperapp.com) automatically retrieves current, past, and forecasted weather data to automatically calculate GDD for plant growth regulator applications. The web-based application automatically suggests the ideal GDD-based reapplication interval based on the PGR selected, application rate, grass species, and area (greens vs. fairways, etc.). GreenKeeper automatically selects the appropriate base temperature depending on the grass species. The number of GDDs that have accumulated, the estimated amount of growth suppression, and the forecasted reapplication date are displayed on the website's homepage. Using this application helps provide uniform growth suppression for a variety of different species and management areas.

Preemergence Herbicide

Q: *After applying a preemergence herbicide, when can I seed turf back into the area?*

A: This is a tricky question to answer because a lot of factors come into play. All preemergence herbicides will inhibit the emergence of all turfgrass seed.

The specific herbicide, rate, and application timing are all important factors in answering how long a standard preemergence herbicide lasts in the soil and when turf can be seeded following an application. Each preemergence herbicide has a different half-life (time required for half the applied herbicide to degrade), which is dependent on soil moisture and temperature. The higher the application rate, the longer the herbicide will take to break down to levels low enough not to impede turfgrass germination.

Lastly, application timing is crucial because the warmer the soil temperatures are following the application, the shorter the life of the herbicide will be in the soil. Having said all of that, it is likely that a three-month waiting period or more (assuming a spring application timing) will be required before you can seed turf following a preemergence herbicide application. See the herbicide label for more information.

Q: *How long can you expect a preemergence herbicide to work?*

A: This is related to the previous question. Generally, a preemergence herbicide will be active in the soil at quantities large enough to suppress germination of crabgrass and other annual grassy weeds for three months or so. However, above-average rainfall in May and June will shorten the longevity of preemergence herbicides due to anaerobic microbial decomposition. Additionally, above-average temperatures will favor weed growth and germination as well as the activity of microbes leading to more crabgrass growth and a shorter herbicide half-life.

Q: *Do preemergence herbicides control broadleaf weeds and sedges, or just grasses?*

A: The following preemergence herbicides inhibit the growth of germinated grasses, which keep seeds from developing (emerging): benefin (Balan), benefin + trifluralin (Team Pro), bensulide (Bensumec), dithiopyr (Dimension), pendimethalin (Pendulum), prodiamine (Barricade), and oxadiazon (Ronstar). These herbicides can inhibit the emergence of broadleaf plants but not as well as grasses.

There are exceptions, of course, with certain preemergence herbicides and specific weeds. Isoxaben (Gallery) inhibits the emergence of broadleaf weeds effectively but does not inhibit the emergence of grasses to the same extent. The preemergence herbicides mentioned in the previous paragraph do not inhibit sedges (that is, yellow nutsedge). However, dimethenamid (Tower) and sulfentrazone (Dismiss, others) do inhibit sedge emergence. Although dimethenamid and sulfentrazone inhibit sedge emergence, herbicides that inhibit sedge emergence typically do not perform as well at preventing grassy weed emergence. For preemergence control of grasses and broadleaf weeds, herbicides such as Crew (dithiopyr + isoxaben) or Gemini (prodiamine + isoxaben) work well.

Q: *I have heard that you can apply a preemergence herbicide in November or December and it will control crabgrass the following year. Is this true?*

A: Late fall applications (especially applications in mid-November or December in the Midwest before the ground freezes) can successfully control crabgrass the following year. The reason for this is that biological (microbial) activity breaks down preemergence herbicides, and since microbial life is most active when soils are warm, the herbicide can be applied to the turf in late fall (and ultimately watered into the soil by irrigation or rainfall) with little risk of any breakdown until the following May when soils warm.

Thus, if your management schedule is very busy in March and early April, you may consider applying a preemergence herbicide in November, December, or February for preemergence crabgrass control instead of the traditional March and April applications. See the herbicide label for more information. For more information about crabgrass control, see *Crabgrass*, page 45.

Q: *Is it OK to aerify turf when you've already applied a preemergence herbicide to control crabgrass?*

A: Crabgrass is a common summer annual weed in managed turf. When you implement good cultural practices but still have some crabgrass, the best approach is to use a preemergence herbicide such as dithiopyr (Dimension), pendimethalin (Pendulum), prodiamine (Barricade), sulfentrazone + prodiamine (Echelon), and others. Since these herbicides work on germinating seeds (prior to emergence, PRE emergence), you must apply them prior to germination. An exception is dithiopyr, which also controls crabgrass after germination until it reaches one tiller.

As a note, the Dimension 2EW label cautions against aerification after a preemergence herbicide application. It recommends to cultivate first. Other preemergence herbicide labels are silent on this topic. All preemergence herbicides bind strongly to the soil and they are not leached by rainfall. Instead, they remain in the top inch or so of soil after application.

Cultivation is the disruptive process of working the soil with limited damage to turf. Aerification is a type of cultivation also called coring, spiking, and punching. The benefits of core aerification (removing cores) are increased water infiltration, increased soil air exchange, decreased thatch, and decreased soil compaction.

In the case of spring core aerification of turf treated with a preemergence herbicide, aerification physically removes plugs of soil containing turf. This process also removes the herbicide and subsequently provides an environment for weed seed to germinate. Keep in mind that a single aerification typically affects only 1%–5% of the total area depending on tine spacing and size. So, if we aerify in the spring, will it negate the preemergence herbicide application and cause crabgrass to go wild? The answer is no, and here's why.

Preemergence herbicides (mitotic inhibitors) work primarily through preventing cell division in the roots of germinating seeds. Symptoms include swelling of root tips and failure of weed seedlings to emerge (develop leaves) following germination. After we aerify, a newly germinating crabgrass seed must still send out a root (the first root is called the radicle), which will come into contact with and take up the preemergence herbicide. When it does, the crabgrass plant will fail to emerge and be controlled. Past research in Michigan and Georgia proves this all to be true.

After aerification, a germinating crabgrass seed could be prevented from emerging by the preemergence herbicide in three different ways (all inside the aerification hole where it is trying to germinate).

1. Crabgrass germinates on the edge of the aerification hole and herbicide uptake occurs from undisturbed soil containing the preemergence herbicide.
2. After aerification, plugs (cores) with herbicide are broken up and returned into the holes.
3. After aerification, the hole refills (fully or partially) from the sides with existing soil containing the preemergence herbicide when soil sloughs off the edges of the aerification channel.

In each case, crabgrass is controlled. Therefore, aerification following a preemergence herbicide application will not result in increased weed emergence or a loss of weed control. Remember that aerification produces a long-term benefit of improving turf health. With improved turf health also comes increased abiotic stress tolerance and increased turf density, which should also help keep out the weeds. Aerification is beneficial and, based on previous research, it should not significantly affect herbicide performance.

Q: How should I control summer annual weeds in spring-seeded, cool-season turf?

A: There are four potential options for controlling weeds in spring-seeded, cool-season turf. The *Herbicides Intended for Use During Spring Seeding* table on page 62 provides additional information.

Herbicides Intended for Use During Spring Seeding

Trade Name (product/A)	Common Name (lb ai/A)	Weed Control Rating[1]				Application Interval		Comments
		Crabgrass	Goosegrass	Yellow Nutsedge	Summer Annual Broadleaves[2]	After Seeding	Before Seeding	
Drive XLR8 (0.7–2.0 qt) Quinclorac 1.5L (0.7–2.0 qt) Quinclorac 75DF (0.33–1.0 lb) QBall (0.7–2.0 qt)	quinclorac (0.25 to 0.75 lb)	PRE:[3] Poor POST: Excellent	PRE: None POST: None	PRE: None POST: None	PRE: Poor POST: Poor to fair	7 days after tall fescue emergence. 28 days after creeping bentgrass, fine fescue, and Kentucky bluegrass	7 days for creeping bentgrass, fine fescue, Kentucky bluegrass. Can be applied day of seeding tall fescue and perennial ryegrass.	Do not use MSO or other adjuvants on new seedlings until at least 28 days after emergence. Excellent control of certain species such as clover. Tank-mixing with Quicksilver can increase spectrum of broadleaf weed control.
Pylex (1.0–1.5 fl oz)	topramezone (0.02–0.03)	PRE: Poor POST: Good	PRE: Poor POST: Excellent	PRE: None POST: None	PRE: Fair POST: Fair	No sooner than 28 days after seeding.	0 days. Can be applied day of seeding tolerant species.	Can be applied to creeping bentgrass.
Quicksilver T&O (1.0–6.7 fl oz) Aim EC (*sod farms only*, 0.5–2.0 oz)	carfentrazone (see label)	PRE: None POST: None	PRE: None POST: None	PRE: None POST: None	PRE: Fair POST: Fair	7 days after emergence.	1 day.	Provides suppression of summer annual broadleaves if applied post-emergence. Does not control grassy weeds.
Tenacity (5–8 fl oz) TRIONE (5–8 fl oz) Meso 4SC Select (5–8 fl oz)	mesotrione (0.156–0.25)	PRE: Fair POST: Good	PRE: Fair POST: Fair	PRE: Fair POST: Fair	PRE: Good POST: Fair	2 mowings or 4 weeks, whichever is longer.	0 days. Can be applied day of seeding tolerant species.	Fine fescues are less tolerant. Tenacity may reduce the density of fine fescue seedings. Do not apply to bentgrass species. An application on the day of seeding followed by an application 4 weeks later is often necessary where crabgrass or goosegrass infestations are severe. Will provide pre- and post emergence control of many grassy and broadleaf weed species.

1 Rating Key: Excellent, >90% control. Good, 75%-90% control. Fair, 50%-75% control. Poor, <50% control.
2 A composite ranking primarily based on control of common purslane and prostrate spurge.
3 PRE = preemergence herbicide timing. POST = postemergence herbicide timing.

1. Apply mesotrione (Tenacity) or topramezone (Pylex) for preemergence control of annual grassy and some broadleaf weeds in newly seeded cool-season turf. These herbicides are more expensive and short-lived but are safe to apply at seeding. Neither work well for establishing creeping bentgrass. The preemergence activity of these herbicides will prevent weeds from germinating in newly planted cool-season turf for about one month.
2. Another strategy is to use a postemergence herbicide instead of a preemergence herbicide to control weeds. Seed in early spring and apply a postemergence herbicide in late May and June; follow label directions. Postemergence herbicide options include carfentrazone (Quicksilver), quinclorac (Drive), mesotrione (Tenacity), and topramezone (Pylex). These products can be safely used soon after seeding to control crabgrass (see label for exact details and safety on each turf species). If the seedlings are more mature (have been mown two or three times following their emergence), then you could use other products such as Q4 Plus (quinclorac + sulfentrazone + 2,4-D + dicamba) or Solitare (quinclorac + sulfentrazone).
3. A third option is to use dithiopyr (Dimension) in May after a spring seeding to control newly germinated crabgrass that has emerged and has not yet tillered. Dithiopyr is the only preemergence herbicide that has postemergence activity on newly germinated crabgrass. This application would also prevent future crabgrass germination through the remainder of summer in newly seeded areas. The assumption with this scenario is that you planted seedlings in early spring and they have developed enough of a root system to tolerate a dithiopyr application and that crabgrass has not yet begun to tiller.
4. The last option is to apply a preemergence herbicide to control crabgrass despite the fact that the lawn may be thin and in need of seeding. Since establishing turf in the spring is not as optimal as establishing turf in the late summer (mid-August to mid-September), you will likely have better long-term results if you wait to seed as opposed to seeding in the spring. This will only be an option for customers patient enough to wait until they get their lawn back into shape. In the meantime, proper fertilization, mowing, and irrigation will encourage recovery.

Picking the Right Formulation

Q: This publication discusses herbicides as standalone products. What about using combination herbicide plus fertilizer products?

A: Granular preemergence herbicides applied on fertilizer carriers are as effective at controlling annual weeds as standalone liquid preemergence herbicide applications. Turf professionals frequently use these combination products.

Postemergence herbicides such as 2,4-D, MCPP, and others are generally much more effective as sprays than when applied as granular products. Sprays are more effective because most postemergence herbicides require foliar uptake and sprayed herbicides are more successful at covering the foliage.

You can improve the foliar uptake of granular products if you apply them when weed foliage is moist (such as in the early morning when dew is present or following irrigation). The moist foliage helps the granules stick to the weed and allows the herbicide to enter the leaf tissue. However, even when applied to moist foliage these products are still generally less effective than sprays at controlling weeds.

Q: Do granular or liquid (sprayed) herbicides work better?

A: See the previous Q&A: *This publication discusses herbicides as standalone products. What about using combination herbicide plus fertilizer products?*, page 63.

Q: Should I use an amine or ester formulation for postemergence broadleaf weed control?

A: Synthetic auxin herbicides (including 2,4-D, 2,4-DP [dichlorprop], dicamba, MCPA, and triclopyr) are commonly formulated in two distinct forms: amine salts and esters.

Amine formulations are generally less volatile than esters and safer to use when applying near sensitive plants (that is, ornamental landscapes and gardens).

Ester formulations have a higher vapor pressure and a much higher volatilization potential.

Research has documented that fall applications with amine or ester formulations of 2,4-D provide optimum broadleaf weed control. However, spring applications are sometimes needed in weedy locations or when you miss an opportunity to make a fall application. Purdue University research demonstrated that during cooler spring months (April), ester formulations are more effective than amine formulations for broadleaf weed control because they more readily penetrate the waxy leaf cuticle. As temperatures in the spring warm, turf managers should switch to amine formulations because they work as well as esters during warmer temperatures in May and early summer. What's more, amine formulations are less volatile than ester formulations, making them safer to use around landscape plantings.

The *Broadleaf Herbicide Formulation* tables on page 64 list products by formulations.

To help track weather conditions and the appropriate time to use amine and ester formulations, visit Michigan State University Extension's GDD Tracker, *https://gddtracker.msu.edu/*.

Q: I have heard that I need to use caution with volatile herbicides. What is herbicide volatility?

A: Volatility is the movement of an herbicide in the air as a gas or vapor. This is different than drift, which is the movement of airborne spray particles during the application. Volatilization occurs after an application when the herbicide volatilizes (or vaporizes) and moves in the air as a gas.

Volatility can increase in hot, dry conditions. The concern with volatilization is that herbicides might move off target and cause damage to desirable ornamentals, fruits, vegetables, and other plants.

Volatility is highly influenced by herbicide selection. Several formulations are referred to as "low volatile" formulations to advertise their reduced risk. The chemical's vapor pressure is a good way to figure out its volatilization potential. As mentioned in the previous Q&A, amines have a lower vapor pressure and thus a lower risk of volatilization than esters. However, within the esters there are different levels of volatility and some esters are formulated to reduce this risk.

Therefore, if you are in a situation when you want to use 2,4-D ester near sensitive plants: (1) consider waiting until fall when ornamentals are less susceptible, and (2) consider switching to an amine if possible. While 2,4-D ester formulations are labeled as "low volatile," this is in contrast to the once used short-chain "high volatile esters," which were sold until the early 1980s. Therefore, all ester formulations are volatile, and the "low volatile" label should not mislead applicators into thinking these are safe to use around sensitive plants.

Emulsion Formulations (Emulsion-in-Water or Microemulsion)	
Ingredients	**Trade Names**
2,4-D choline + fluroxypyr + halauxifen	GameOn
2,4-D + fluroxypyr + dicamba	Escalade 2
2,4-D + MCPP + dicamba + carfentrazone	SpeedZone EW
2,4-D + MCPP + dicamba + carfentrazone	SpeedZone Southern EW

Spray Water

Q: Does the pH and hardness of my spray water influence weed control?

A: The quality of the water in herbicide applications is often overlooked. Research clearly shows that the quality of water used for spraying can affect how herbicides perform. Both water pH and hardness can affect herbicide performance. Hardness level is most critical.

Recent Purdue research demonstrates that hard water can reduce the efficacy of weak acid herbicides. Commonly used weak acid herbicides include 2,4-D, 2,4-DP, clopyralid, dicamba, glyphosate, MCPA, mecoprop (MCPP), mesotrione, quinclorac, and triclopyr. Our research has focused on how hard water can reduce the efficacy of 2,4-D on broadleaf weeds. Current research has shown that hard water antagonizes amine formulations but does not antagonize ester or emulsified formulations (see *Broadleaf Herbicide Formulation* tables on this page). Hard water antagonism of broadleaf herbicides occurs when water hardness levels approach or exceed 400 ppm, which is common in the Midwest.

It is possible to overcome hard water antagonism by adding ammonium sulfate or a water conditioning agent to the spray tank. Typically, water conditioning agents (conditioners) contain ammonium sulfate, which counteracts the effects of the hard water. Mix 8.5 to 17 lb of sprayable dry ammonium sulfate (21-0-0) per 100 gallons of spray water or add the liquid water conditioning agent of your choice into your spray tank water before adding the herbicide. The spray grade ammonium sulfate is water-soluble and will dissolve in the tank.

The Impact of Water Quality on Pesticide Performance (Purdue Extension publication PPP-86) describes how to test your water quality for better performance of your herbicides. It is available from the Education Store, *https://.edustore.purdue.edu.*

Q: What are surfactants and adjuvants, and how much do I add to my spray tank when mixing?

A: There are many terms when discussing herbicide additives, including wetting agents, spreaders, stickers, surfactants, nonionic, methylated seed oils, crop oils, and more. These terms

Amine Formulations (Water Soluble)	
Ingredients[1]	**Trade Names**
2,4-D	2,4-D Amine 4, Clean Amine, Saber, Opti-Amine, Shredder Amine 4, WEEDestroy AM-40, Weedar 64
2,4-D + clopyralid + dicamba	Millennium Ultra 2
2,4-D + dicamba + sulfentrazone + penoxsulam	Avenue South
2,4-D + MCPA + triclopyr + dicamba	Triad T Select
2,4-D + MCPA + dicamba	Triad Select
2,4-D + MCPA + dicamba + sulfentrazone	Triad SFZ Select
2,4-D + MCPP + dicamba	3-D, Eliminate LO, Eliminate-D, End-Run, MEC Amine-D, Strike 3, Threesome, Three-Way, Trimec 1000, Trimec 992, Trimec Bentgrass Formula, Trimec Classic, Trimec LAF-637, Triplet, Trimec Southern, Triplet Low Odor, Triplet SF, TruPower2, TruPower3, Vessel
2,4-D + MCPP + dicamba + sulfentrazone	Surge, SureZone
2,4-D + MCPP + 2,4-DP	Spoiler, Triamine
2,4-D + quinclorac + dicamba	2DQ, Momentum Q, Quincept
2,4-D + quinclorac + dicamba + sulfentrazone	Q4 Plus
2,4-D + triclopyr	Aquasweep, Chaser 2 Amine, Turflon II Amine
2,4-D + triclopyr + dicamba + sulfentrazone	Foundation
2,4-D + triclopyr + fluroxypyr + sulfentrazone	Momentum 4-Score
dicamba	Banvel, Clarity, Clash, Cruise Control, Detonate, Diablo, Rifle, Sterling Blue, Strut, Topeka, Vanquish, Vision
MCPA	MCPA-4 AMINE
MCPA + MCPP + dicamba	Trimec Encore, Tri-Power
MCPA + triclopyr + dicamba	Eliminate, Horsepower

1 Some products are formulated with a mixture of amines and esters, including Battleship III (MCPA + fluroxypyr + triclopyr), Change Up (MCPA + fluroxypyr + dicamba), Chaser Ultra 2 (MCPA + fluroxypyr + 2,4-DP), Momentum FX2 (2,4-D + triclopyr + fluroxypyr), and Tailspin (triclopyr + fluroxypyr). In these specific examples, fluroxypyr is an ester; the 2,4-D, 2,4-DP (dichlorprop), MCPA, and triclopyr are amines.

Ester Formulations (Water Insoluble, Emulsifiable)	
Ingredients	**Trade Names**
2,4-D	Barrage HF, Shredder 2,4-D LV4, Shredder 2,4-D LV6, Shredder E-99, Weedone LV4 EC
2,4-D + 2,4-DP	Patron 170
2,4-D + 2,4-DP + dicamba	Brushmaster, Super Trimec
2,4-D + MCPP + dicamba + pyraflufen	RedZone 2
2,4-D + triclopyr	Chaser, Crossroad, CrossCut Select, Everett
2,4-D + triclopyr + dicamba + pyraflufen	4-Speed XT
2,4-D + triclopyr + dicamba + sulfentrazone	TZONE SE
2,4-D + triclopyr + fluroxypyr + flumioxazin	Sure Power
fluroxypyr	Vista XRT
MCPA	MCPA ESTER 4, Wildcard
MCPA + MCPP + dicamba + carfentrazone	PowerZone
MCPA + triclopyr + dicamba	Cool Power, Lesco ThreeWay Ester II, Spurge Power
triclopyr	Boulder, Turflon Ester Ultra, Triclopyr 4

can be confusing. For more information, see *Adjuvants and the Power of the Spray Droplet: Improving the Performance of Pesticide Applications* (Purdue Extension publication PPP-107), available from the Education Store, https://.edustore.purdue.edu. *Compendium of Herbicide Adjuvants* (Purdue Extension publication PPP-115 and WS-54) lists adjuvant products and their properties. It is available from the Education Store or at https://edustore.purdue.edu/ppp-115.html.

Many herbicide labels recommend adding a surfactant to improve herbicide performance. A few herbicides state that surfactants can increase turfgrass injury in some conditions such as drought or high temperatures. If the label encourages the use of a surfactant, then use one to improve weed control. In some cases, the manufacturer may make surfactant use optional when it is hot or there are stressful conditions. Some labels state not to add surfactants.

The most common surfactant for turf herbicides is a nonionic surfactant (NIS). Methylated seed oils (MSO) and crop oil concentrates (COC) are also recommended for some herbicides. Use the specific type of surfactant recommended on the herbicide label and use the recommended rate. Your pesticide supplier should also have the types of surfactants you need.

Q: *What should my herbicide spray volume be per acre or per 1,000 square feet?*

A: Most herbicide labels specify a range of spray volumes or a minimum spray volume of water per acre. In general, herbicides are more effective at lower spray volumes than insecticides or fungicides, but weed control may be reduced if too low of a spray volume is used (which may result in poor spray coverage).

For many years, herbicide applications to turf were made at large spray volumes (≥87 gal/A, ≥2 gal/1,000 ft^2) because of the type of equipment applicators used (lawn hand gun, large boom sprayers). However, the increased use of low-volume (<20 gal/A, often 12–15 gal/A) spray equipment in lawns (ride-on sprayers/spreaders) and the desire to reduce transportation costs means that many applicators are now spraying as little as 11 gal/A (0.25 gal/1,000 ft^2).

With these reduced application volumes comes decreased spray coverage (depending on nozzle type) and possibly reduced weed control. For most herbicides, it is ideal to use a spray volume of 30 gal/A (0.7 gal/1,000 ft^2) or more. Most herbicides recommend a minimum carrier volume of water for spray applications. This minimum is typically 20 gal/A, but it varies by herbicide. Some herbicides say very little about spray volume. Some herbicide labels such as the Q4 Plus (quinclorac + sulfentrazone + 2,4-D + dicamba) label even provide specific warnings about low-volume applications.

We are currently researching the use of low spray volumes for weed control. To date, our research demonstrates that applying herbicides with low-volume spray equipment is as effective as using traditional high-volume spray equipment.

Warm-Season Turfgrasses

Q: *Can I safely spray bermudagrass and zoysiagrass with a nonselective herbicide such as glyphosate to control weeds during winter?*

A: Bermudagrass and zoysiagrass enter winter dormancy after the first frost each fall and remain dormant during winter. Winter annual weeds invade during these winter months. Because new spring growth of bermudagrass and zoysiagrass originates from the crowns, the brown dormant grass leaves during winter are dead and will not move herbicide to the stem. Thus, you can apply a nonselective herbicide during winter to kill winter weeds without injuring the bermudagrass or zoysiagrass. However, there are some precautions you must follow to be successful.

First, don't wait too long because an application will injure any new growth of these grasses in early spring. Even if only 10% green turf is visible, you could delay full spring green-up by two or more weeks if you apply a nonselective herbicide at that time.

Second, even if the turf appears dormant on top, further examination may reveal new leaf growth and green stolons. Inspect the turf carefully to make sure there is no new growth prior to applying a nonselective herbicide to bermudagrass or zoysiagrass during winter.

Third, when applying a nonselective herbicide to dormant bermudagrass or zoysiagrass during winter, do not use spot treatments. Also, do not apply with a boom sprayer with more than 20 gallons per acre carrier volume because this will increase the risk of damaging crowns, stolons, or new growth.

Q: *How can I safely remove overseeded turf (that is, perennial ryegrass) without injuring the bermudagrass?*

A: Many sulfonylurea herbicides are labeled for the selective removal of cool-season grasses such as perennial ryegrass from bermudagrass. You can safely use these products in dormant bermudagrass or during spring green-up when bermudagrass is emerging from winter dormancy. Not all sulfonylurea herbicides are useful for removing cool-season grasses from a warm-season turf; however, the following products are labeled for this use: flazasulfuron (Katana), foramsulfuron (Revolver), metsulfuron (Manor, Mansion, MSM), rimsulfuron (Rimsulfuron 25DF), sulfosulfuron (Certainty), and trifloxysulfuron (Monument, Recognition).

Flazasulfuron, rimsulfuron, and trifloxysulfuron are the sulfonylureas that generally control perennial ryegrass more quickly (one or two weeks). Metsulfuron, foramsulfuron, and sulfosulfuron may take two to four weeks to control perennial ryegrass. Pronamide (Kerb) is another herbicide that is not a sulfonylurea but can be used to slowly remove perennial ryegrass from bermudagrass—usually less than four weeks.

How quickly these products control perennial ryegrass and other weeds depends mostly on temperature. Researchers have shown about a 1% increase in control for each degree (F) increase in soil temperature. All of these sulfonylurea herbicides are applied at low use rates and most require a nonionic surfactant. Many of the sulfonylurea herbicides control selected winter annual grassy and broadleaf weeds (such as annual bluegrass and common chickweed) in addition to perennial ryegrass. Most of these herbicides require a second application for complete control. For exact rates and more information, consult the herbicide label.

Q: Can a preemergence herbicide be used when planting bermudagrass or zoysiagrass?

A: Bermudagrass and zoysiagrass are warm-season turfgrasses that can be established by seed, sod, plugs, or sprigs (pieces of stolons and rhizomes). There is no safe preemergence herbicide that can be used at the time of planting bermudagrass or zoysiagrass seed. When establishing bermudagrass or zoysiagrass from sprigs or sod, most preemergence herbicides, such as dithiopyr, pendimethalin, and prodiamine, will prevent the desirable grasses from producing healthy roots in the same way that they prevent weeds from developing healthy roots. However, there is one herbicide that can be applied at the time of planting that will allow for normal healthy bermudagrass and zoysiagrass root development while still preventing the emergence of problematic summer annual grasses. That herbicide is oxadiazon. Oxadiazon is sold in both a liquid formulation and a water soluble powder (WSP) formulation through both the trade name product Ronstar as well as post-patent herbicides. A formulation of oxadiazon called Ronstar G was also available until 2020 but its production was phased out. Some existing quantities may still be available. An alternative granular solution is oxadiazon on fertilizer carrier. This common combination product should be sold by distributors in areas where warm-season turfgrasses are grown. See the labels of these products for more detailed application instructions.

Miscellaneous

Q: I have some product in my pesticide storage room that is several years old. Is it still good?

A: First, you should clearly date all product containers as they are delivered. If the liquid product has been stored in an area that is not exposed to extreme temperatures (too hot or too cold—not frozen) and there is no physical separation of the contents, then it is likely still good for use—assuming the product is still registered for use in your state and that you follow the label directions. However, it would be wise to test a small weedy area of turf to make sure the herbicide still works before you make the application.

If the product has been exposed to extreme temperatures, the contents have settled, you are unsure about the contents of the container, the pesticide label is no longer affixed to the container, or it is no longer registered for use in your state, then dispose of it properly.

An Indiana Pesticide Clean Sweep Project (sponsored by the Office of Indiana State Chemist [(OISC)] collects and disposes of suspended, canceled, banned, unusable, opened, unopened, or just unwanted pesticides (weed killers, insecticides, rodenticides, fungicides, miticides, etc.). To learn about the next drop-off date, call the OISC at (765) 494-1492. If you live outside of Indiana, contact your state regulator (see page 124) to learn about pesticide disposal programs.

Pesticides and Their Proper Storage (Purdue Extension publication PPP-26) contains a pesticide checklist to determine whether your pesticides are stored safely and correctly. It is available from the Purdue Extension Education Store, https://.edustore.purdue.edu.

Q: I purchased some herbicide a few years ago and still have some in stock. However, since then the label uses for this herbicide have changed. Can I still apply it to turf?

A: It depends. To help explain, here are two examples.

MSMA is no longer registered for use in home lawns. Thus, if you purchase a new 2.5-gallon container of MSMA it can only be used on sod farms, highway rights-of-way, and golf courses on turf. However, if you purchased MSMA prior to December 31, 2010, your container might have a label that states it can be used on lawns and not just sod farms, highway rights-of-way, and golf courses. Thus, whether or not you can treat a lawn legally with this product depends on the language of the label on the container. See *Can I still use MSMA?* (below) for details.

Similarly, Certainty (sulfosulfuron) herbicide purchased before May 2011 is allowed for use on cool-season species. The current Certainty label does not allow use on cool-season species. However, the herbicide Sertay (sulfosulfuron) with the same ingredient currently allows use on some cool-season species. Thus, the areas you can legally treat with Certainty depend on the language of the label affixed to the container.

Q: Can I still use MSMA?

A: The use of MSMA on residential turf (which includes parks and athletic fields) and on commercial turf ended on December 31, 2010. Remaining MSMA supplies can be applied to those areas until they are exhausted.

For sod farms, golf courses, and highway rights-of-way, the following dates and application restrictions apply:

- MSMA sold after December 31, 2010, has new labeling restrictions for golf course and sod farm use. Those restrictions are as follows
  - Golf courses: One broadcast application will be allowed on newly constructed courses. Application on existing courses will be limited to spot treatment (100 ft^2 per spot), not to exceed 25% of the total course in one year.
  - Sod farms: Two broadcast applications will be allowed per crop. A 25-foot buffer strip will be required for fields bordering permanent water bodies.
  - Highway rights-of-way: Two broadcast applications will be allowed per year. A 100-foot buffer strip will be required adjacent to permanent water bodies.
- The Organic Arsenical Products Task Force (OAPTF, https://www.oaptf.com) announced that use of MSMA for weed management on golf courses, sod farms, and highway rights-of-way will continue beyond the original phase-out date of June 30, 2013. The National Academy of Sciences is conducting a review regarding the risk of inorganic arsenic. This review is expected to be completed in 2017. Thus, the MSMA use on these areas will be extended until the scientific review is completed and the EPA finalizes their registration review. The EPA registration review was scheduled for completion in 2019 but an update on the future registration of this herbicide is still forthcoming.

How to Use the Tables in this Publication

- Herbicides included in this publication are registered for use in most states with a few exceptions (see below). Although products listed in this guide are registered for use, not all products may be available from local distributors.
- The tables are grouped in nine sections:
  1. Nonselective herbicides/fumigants for turfgrass renovation
  2. Nonselective herbicides for turfgrass border maintenance (edging)
  3. Preemergence herbicides
  4. Postemergence herbicides
  5. Sedge control herbicides
  6. Preemergence weed control in creeping bentgrass putting greens
  7. Postemergence weed control in putting greens
  8. Plant growth regulators for general turf use
  9. Annual bluegrass suppression in creeping bentgrass putting greens with plant growth regulators (PGR)
- The preemergence, postemergence, and sedge control sections also include other tables. The first summarizes how well each herbicide controls specific weeds. The second summarizes label information about the turfgrass species that can be treated. Weed efficacy information is based upon Purdue University research, label information, and published data from other weed scientists.
- In each section a table will present information similar to the sample table below. These tables are guides and are not substitutions for herbicide or PGR labels.
- **Column 1—Trade Name** (product/A)—provides the trade name for a product or products with the same common active ingredient. In parentheses next to each name is range of use rates (rate per acre) specified on the label.
- **Column 2—Common Name** (lb ai/A)—provides the herbicide common name (also known as active ingredient). In parentheses is the amount of active ingredient (ai) or acid equivalent (ae) applied per acre according to the label rates provided in Column 1.
- **Column 3—Weeds Controlled**—provides a summary of the type of weeds controlled. *This is not an exhaustive list but merely a summary of some specific weeds or types of weeds. See labels for complete lists of weeds controlled.*
- **Column 4—Turfgrasses**—lists common turfgrass species that can be safely treated per label instructions.
- **Column 5—Comments**—provides important considerations when using this product. *This is not a full list of precautions or use instructions and is meant only to point out a few key items. The label contains a full list of restrictions and precautions.*
- *Always, read the label and refer to the label for the most up-to-date information about these products before selecting or using them.*
- Products that are labeled for use in only warm-season grasses (bermudagrass, buffalograss, and zoysiagrass) or predominantly warm-season turf are indicated with this symbol: (W). Do not use these products on cool-season turf except by label directions.
- Products with sale and/or use restrictions in New York State or Long Island, New York (Nassau and Suffolk counties), are indicated with this symbol: (NY). Refer to the label for more information.
- Products registered only in specific states (primarily those states with significant bermudagrass turf) are indicated with this symbol: (L). Check with your state regulatory agency and local suppliers for availability.

Sample Table

Trade Name (product/A)	Common Name (lb ai/A)	Weeds Controlled	Turfgrasses	Comments
Celsius WG (2.5–4.9 oz) (W) (NY)	thiencarbazone + iodosulfuron + dicamba (see label)	broadleaf weeds and grasses	bermudagrass, buffalograss, zoysiagrass	**Do not use on cool-season turf.** This product provides effective control of many broadleaves and grassy weeds in bermudagrass, buffalograss, and zoysiagrass. Do not use more than 7.4 oz/A per year.
Drive XLR8 (0.7–2.0 qt) Quinclorac 1.5L (0.7–2.0 qt) Quinclorac 75DF (0.33–1.0 lb) others (NY)	quinclorac (0.25–0.75)	crabgrass, barnyardgrass, foxtail, white clover, dandelion	annual bluegrass, bermudagrass, creeping bentgrass, fine fescues, Kentucky bluegrass, perennial ryegrass, tall fescue, zoysiagrass	Not labeled for golf course putting greens or collars. Use lower rates on creeping bentgrass and fine fescues. Add COC (2 pt/A) or MSO (1.5 pt) to increase performance. Generally safe to use on newly seeded turf, although adding surfactant is not recommended on new seedlings. See label for details. Momentum Q, Q4 Plus, Quincept, and Solitare also contain quinclorac.

(W) This product is labeled for use primarily in warm-season turf. Read the label for a list of tolerant grasses.
(NY) Use or sale of this product has restrictions in New York state or Long Island, New York (Nassau and Suffolk counties). Refer to the label for more information.

Nonselective Herbicides/Fumigants for Turfgrass Renovation

Refer to herbicide labels for specific use instructions. Products containing glyphosate are listed from most concentrated to least concentrated formulations.

Trade Name (product/A)	Common Name (ingredient rate)	Weeds Controlled	Use	Comments
Basamid G (135–525 lb)	dazomet (1.35–5.25 lb ai/A)	most weeds, nematodes, and soil diseases	soil fumigant for renovation	**Restricted use pesticide.** The soil should be dry prior to application. Soil temperatures must be 43–90°F. Apply with a drop spreader. See label for instructions on methods of preparation and application. The soil should be kept uniformly moist but not waterlogged for 72 hours after application. See label for additional instructions. See the label for PPE requirements and reentry restrictions.
Campaign (1.5–4.0 pt) NY	glyphosate + 2,4-D (see label)	most weeds	dormant bermudagrass; controls existing vegetation	Apply to areas that are being renovated before establishment of the desired turfgrass or to dormant bermudagrass. Multiple applications may be needed to control certain perennial weeds.
Refuge (0.33–3.0 qt)	glyphosate[1] (6.22 lb ai/gal; 5.0 lb ae/gal)	most weeds	controls existing vegetation	Apply to areas that are being renovated before establishment of the desired turfgrass. Multiple applications may be needed to control certain perennial weeds. Wait for regrowth before making the follow-up applications. Aqua Neat and Rodeo are labeled for aquatic use, so these products can be used to control vegetation near aquatic areas.
Roundup PROMAX (1.0–3.0 qt) Roundup Original MAX (0.4–3.3 qs) others	glyphosate[1] (5.5 lb ai/gal; 4.5 lb ae/gal)			
Aqua Neat (1.5–7.5 pt) Rodeo (1.5–7.5 pt)	glyphosate[1] (5.4 lb ai/gal; 4.0 lb ae/gal)			
Departure (0.4–3.6 qt) Prosecutor (0.4–3.6 qt) Touchdown Total (0.4–3.6 qt)	glyphosate[1] (4.17 lb ae/gal)			
Roundup PRO (0.75–5.0 qt) Razor Pro (0.4–5.0 qt) others	glyphosate[1] (4.0 lb ai/gal; 3.0 lb ae/gal)			

1 Some glyphosate formulations provide improved weed control when you mix in a water conditioner (to correct water hardness). You may also need to add a surfactant. See label for instructions. For more information see *The Impact of Water Quality on Pesticide Performance* (Purdue Extension publication PPP-86), available from the Education Store, https://edustore.purdue.edu, (765) 494-6794.

NY Use or sale of this product has restrictions in New York state or Long Island, New York (Nassau and Suffolk counties). Refer to the label for more information.

Nonselective Herbicides for Turfgrass Border Maintenance (Edging) or Dormant Warm-Season Turfgrass Applications

Refer to herbicide labels for specific use instructions.

Trade Name (product/A)	Common Name (lb ai/A)	Weeds Controlled	Use	Comments
Barrier Year-Long Vegetation Control Concentrate (6.12 gal/A)	2,4-D + imazapyr + fluazifop + dicamba + pyraflufen-ethyl	existing vegetation and preemergence control of weeds	controls existing vegetation	This is a nonselective herbicide for controlling vegetation (weeds and desirable grasses) and providing residual weed control for up to 12 months in the treated area. Do not use in areas intended to be seeded or planted in the next 12 months. Read label for information on injury to nontarget plants.
Finale (2–6 qt) NY	glufosinate (0.5–1.5) 11.33% or 1.0 lb ai/gal formulation	existing vegetation	controls existing vegetation	Use 2–4 fl oz/gal of water for spot applications and 2–6 qt/A for broadcast applications. Finale has limited translocation. Adding 8.5 lb of ammonium sulfate (spray grade) per 100 gal of water may improve the level of weed control.
Finale XL T&O (27–82 fl oz) Cheetah Pro (27–82 fl oz) Surmise Pro (48–80 fl oz) NY	glufosinate (0.5–1.5) 24.5% or 2.34 lb ai/gal formulation	existing vegetation	controls existing vegetation	Use 1.7–3.2 fl oz/gal of water for spot applications and (27–82 fl oz/A for broadcast applications. Finale XL T&O has limited translocation. It will leave a straight line when edging turfgrasses. Adding 8.5 lb of ammonium sulfate (spray grade) per 100 gal of water may improve the level of weed control. This herbicide may be used to control winter annual weeds in established dormant bermudagrass.
ProDeuce (4.1–8.2 qt)	glyphosate + prodiamine (4.1 + 0.75–8.2 + 1.5)	existing vegetation and preemergence control of weeds	controls existing vegetation	Mix 3–6 oz in 1 gal of water and treat 1,000 ft^2. For best results, apply on a sunny day when the temperature is above 60°F and the air is calm. Do not use in lawns or for lawn renovations as this product prevents desirable grass establishment.
QuickPro (2.25–12.25 lb) Razor Burn (7.5 qt)	glyphosate + diquat (see label)	existing vegetation	controls existing vegetation	For spot treatments add 1.2 oz/gal for easy-to-kill annuals, or up to 1.5 oz/gal for perennial weeds. Spray to wet; do not spray to runoff. Provides quicker burn than glyphosate alone. Do not apply more than 12.25 lb/A/year of QuickPro.
Reward (1–2 pt) Diquat SPC 2L (1–2 pt) others	diquat (0.25–0.5)	existing vegetation	controls existing vegetation	For spot spray, add 0.75 fl oz/gal. Add NIS at 0.25% (v/v) to the spray mix. Use at least 15 gal/A spray volume. Increased water volumes (60–100 gal/A) will enhance activity. Because it is not translocated, it will leave a straight line when edging creeping turfgrasses. However, weed control is not as effective as glyphosate.
Roundup Pro (see label) See *Nonselective Herbicides/Fumigants for Turfgrass Renovation* (page 68) for a more detailed list of products with glyphosate	glyphosate (see label)	existing vegetation	controls existing vegetation	Use as a spot treatment to prevent turfgrass encroachment in fences, paved areas, walkways, and around ornamental plantings. For best results, apply on a sunny day when the temperature is above 60°F and the air is calm. Spray to wet; do not spray to runoff.
Scythe (see label)	pelargonic acid (see label)	existing vegetation	controls existing vegetation	This product is a contact herbicide, which is nonselective. It is not translocated and will only control plant parts that are sprayed. This product is quick-acting and effects are seen within hours. This is an organic product made of fatty acids.
Surmise SpeedPro XT (18.75–37.5 fl oz/A) Onmark (18.75–37.5 fl oz/A) W NY	glufosinate + pelargonic acid + imazethapyr	existing vegetation and preemergence control of weeds	controls existing vegetation	Mix at 2.5% to 5% for spot treatments to undesirable vegetation. This herbicide has limited translocation. Avoid applications to wet plants or if rainfall is expected within 2 hours. Consult the label regarding drift precautions. In residential lawns, use only spot treatments. See page 65 for tips on nonselective herbicide use in dormant warm-season grasses.

W This product is labeled for use primarily in warm-season turf. Read the label for a list of tolerant grasses.

NY Use or sale of this product has restrictions in New York state or Long Island, New York (Nassau and Suffolk counties). Refer to the label for more information.

Preemergence Herbicides

Weed Control Ratings for Preemergence Herbicides

Herbicide	annual bluegrass	barnyardgrass	bittercress	chickweed, common	crabgrass	goosegrass	henbit	knotweed, prostrate	purple deadnettle	purslane	speedwell, corn	spurge, prostrate	yellow foxtail	yellow woodsorrel
atrazine (Aatrex)	E	P	E	E	F	P	E		E	G	E	G	P	G
benefin (Balan)	G	G	P	G	G	F	G	P	G		P	P	G	
benefin + oryzalin (Surflan XL)	G	G	P	G	E	G	G	G	G	G		F	G	E
benefin + trifluralin (Team Pro)	G	G		G	G	F	G		G			F	G	E
bensulide (Bensumec, PreSan)	F	G	P	P	G	F	P		P	F			G	
bensulide + oxadiazon (Goosegrass/Crabgrass)	G	G			G	G							G	
dimethenamid (Tower)	F		G		F	G				P		G		
dimethenamid + pendimethalin (FreeHand)	G	G	G	G	E	G	G	G	G	P	F	G	G	E
dithiopyr (Dimension)	G	G	G	G	E	G	G	G	G	P	G	G	G	G
dithiopyr + isoxaben (Crew)	G	G	E	E	E	G	E	E	E	E	G	E	G	G
ethofumesate (Prograss)	G	F		F	F					F			F	
flumioxazin (Stayguard, SureGuard)	E	G	G	G	G	G	G			G		G	G	G
indaziflam (Specticle FLO or G)	E	G	G	G	E	E	G	G		G	G	G	G	
isoxaben (Gallery)	P	P	E	E	P	P	E	E	E	E	G	E	P	G
methiozolin (PoaCure)	E		G	G	G	G					G			
metolachlor (Pennant MAGNUM)	G	G		F	G	P				F		P	G	
metribuzin (Sencor)	E	P	E	E	F	P	E		E	G	E	G	P	G
oryzalin (Surflan)	E	G	P	G	E	G	G	G	G	G	P	F	G	
oxadiazon (Ronstar)	G	G	P	P	G	E	P	P	P	F	G	F	G	F
pendimethalin (Pendulum)	G	G	G	G	E	G	G	G	G	P	F	G	G	E
prodiamine (Barricade)	G	G	G	G	E	G	G	G	G	P	G	G	G	F
prodiamine + isoxaben (Gemini)	G	G	E	E	E	G	E	E	E	E	G	E	G	G
prodiamine + quinclorac (Cavalcade PQ)	G	G	G	G	E	G	G	G	G	P	G	G	G	F
prodiamine + sulfentrazone (Echelon)	G	G	G	G	E	G	G	G	G	P	G	G	G	
pronamide (Kerb)	E	G		E	F	F	F	G	F	G	E	P	G	
simazine + prodiamine + imazaquin (Coastal)	E	G	E	E	E	G	E	G	E	E	E	G	G	F
simazine (Princep)	E	G	E	E	F	P	E		E	E	E	F	G	

Rating Key: E = excellent (≥90% control). G = good (75%–90% control). F = fair (50%–75% control). P = poor control (≤50% control). Blank = no data.

Turfgrass Tolerance to Preemergence Herbicides

Herbicide	Cool-Season						Warm-Season		
	annual bluegrass	creeping bentgrass	fine fescue	Kentucky bluegrass	perennial ryegrass	tall fescue	bermudagrass	buffalograss	zoysiagrass
atrazine (Aatrex)	NR	NR	NR	NR	NR	NR	NR	NR	S-I
benefin (Balan)	I-S	S	S	S	S	S	S	S	S
benefin + oryzalin (Surflan XL)	NR	NR	NR	NR	NR	S-I	S	S	S
benefin + trifluralin (Team Pro)	NR	S	S	S	S	S	S	S	S
bensulide (Bensumec)	NR	S	S	S	S	S	S	S	S
bensulide + oxadiazon (Goosegrass/ Crabgrass)	NR	S	NR	S	S	S	S	NR	S
dimethenamid (Tower)	NR[1]	NR[1]	NR[1]	NR[1]	NR[1]	NR[1]	S	S	S
dimethenamid + pendimethalin (FreeHand)	NR	NR	NR	NR	NR	NR	S	S	S
dithiopyr (Dimension)	NR	S	S-I	S	S	S	S	S	S
dithiopyr + isoxaben (Crew)	NR	S	S	S	S	S	S	S	S
ethofumesate (Prograss)	NR	S	NR	S	S	S	S	NR	NR
flumioxazin (StayGuard, SureGuard)	NR	NR	NR	NR	NR	NR	S[2]	NR	S[2]
indaziflam (Specticle FLO or G)	NR	NR	NR	NR	NR	NR	S	S	S
isoxaben (Gallery)	NR	S	S	S	S	S	S	S	S
methiozolin (PoaCure)	NR	S	S	S	S	S	S	NR	S
metolachlor (Pennant MAGNUM)	NR	NR	NR	NR	NR	NR	S	NR	S
metribuzin (Sencor)	NR	NR	NR	NR	NR	NR	S	NR	NR
oryzalin (Surflan)	NR	NR	NR	NR	NR	S-I	S	S	S
oxadiazon (Ronstar)	NR	S[3]	S[3]	S[3]	S[3]	S[3]	S	S	S
pendimethalin (Pendulum)	S[4]	S[4]	S	S	S	S	S	S	S
prodiamine (Barricade)	NR	S	S	S	S	S	S	S	S
prodiamine + isoxaben (Gemini)	NR	S	S	S	S	S	S	S	S
prodiamine + quinclorac (Cavalcade PQ)	NR	S-I	S-I	S	S	S	S	S	S
pronamide (Kerb)	NR	NR	NR	NR	NR	NR	S	S	S
simazine (Princep)	NR	NR	NR	NR	NR	NR	S	S	S
simazine + prodiamine + imazaquin (Coastal)	NR	NR	NR	NR	NR	NR	S[2]	NR	S[2]
sulfentrazone + prodiamine (Echelon)	NR	S	S	S	S	S-I	S	S	S

Rating Key: S = safe at labeled rates on healthy, mature turf. I = intermediate safety or some injury may occur; may cause minor damage to mature, healthy turf. Consider using the lower end of the rate range. Do not apply to turf under stress. NR = not registered for use on this species.

1 For use on cool- and warm-season grasses on golf courses (see label), and warm-season use only on sod farms, commercial and recreational turf, and residential turf.

2 Can be used safely on dormant bermudagrass or zoysiagrass.

3 Oxadiazon 2G is the only oxadiazon formulation registered for use on cool-season turf.

4 Pendulum 3.3EC and other 3.3EC formulations of pendimethalin are not labeled for use on creeping bentgrass or annual bluegrass.

Preemergence Herbicides

Refer to labels for weeds controlled and specific turfgrass species use instructions.

Trade Name (product/A)	Common Name (lb ai/A)	Weeds Controlled	Turfgrasses	Comments
AAtrex 4L (1–2 qt) AAtrex Nine-O (1.1–2.2 lb) others W NY	atrazine (1.0–2.0)	summer annual grasses, annual bluegrass, henbit, common chickweed, some small-seeded broadleaves in warm-season turf	established bermudagrass, zoysiagrass	**Do not use on cool-season turf.** Restricted use pesticide due to ground and surface water concerns. Apply between Nov. 1 and Dec. 15 for winter weed control. Apply atrazine only to dormant turf. Do not apply to areas that drain into desirable cool-season turfgrasses. Avoid applying during spring green-up. Do not apply more than 2.0 lb ai/A for any application and 1.0 lb ai/A of the sprayable formulation in residential turf. Do not apply more than 3.0 lb ai/A per year.
Balan 2.5 G (60–120 lb) Crabgrass Preventer with 2.5% Balan (60–120 lb) L	benefin (1.5–3.0)	summer annual grasses, annual bluegrass, some small-seeded broadleaves	established bermudagrass, creeping bentgrass, fine fescue, Kentucky bluegrass, perennial ryegrass, tall fescue, zoysiagrass	Do not use on golf course putting greens. May thin established annual bluegrass. For annual bluegrass control, apply 1–2 weeks before expected germination. In areas of heavy annual bluegrass infestation, control will result in temporary thin areas.
Barricade 4FL (0.625–3.0 pt) Prodiamine 4L (0.625–3.0 pt) Barricade 65 WG (0.5–2.3 lb) Prodiamine 65 WDG (0.5–2.3 lb) Barricade and prodiamine G available on fertilizer carrier (see label) others	prodiamine (0.31–1.5)	summer annual grasses, annual bluegrass, henbit, common chickweed, spurge, some small-seeded broadleaves	established bermudagrass, buffalograss, creeping bentgrass, fine fescue, Kentucky bluegrass, perennial ryegrass, tall fescue, zoysiagrass	Apply only to well-established turf. Do not apply to putting greens or creeping bentgrass maintained < 0.5 inch. Do not exceed 1.5 lb ai/A per year on warm-season turf or tall fescue. Do not exceed 0.94 lb ai/A per year on Kentucky bluegrass or perennial ryegrass. Do not exceed 0.66 lb ai/A per year on creeping bentgrass. Do not make more than 2 applications per calendar year, and allow at least 60 days between treatments. Wait 4–6 months before reseeding following an application.
Bensumec 4LF (1.88–3.13 gal) Pre-San 12.5G (80–100 lb) Weedgrass Preventer (146 lb)	bensulide (see label)	controls annual grasses and selected broadleaf weeds	established bermudagrass, buffalograss, creeping bentgrass, fine fescue, Kentucky bluegrass, perennial ryegrass, tall fescue, zoysiagrass	Apply in the spring for preemergence control of annual grassy weeds such as crabgrass and goosegrass. Apply high rate in late summer for annual bluegrass control. Irrigate immediately after treatment. Delay reseeding for 4 months after treatment. Limit of 2 applications per year on putting greens. Do not use on putting greens composed of more than 50% annual bluegrass (*Poa annua*).
Cavalcade PQ (1.0–2.3 lb) NY	prodiamine + quinclorac (see label)	summer annual grasses, annual bluegrass, henbit, common chickweed, spurge, some small-seeded broadleaves	established bermudagrass, buffalograss, creeping bentgrass, Kentucky bluegrass, perennial ryegrass, tall fescue, zoysiagrass	Apply only to well-established turf. Do not apply to golf course putting greens or collars. Do not exceed 4.6 lb/A per year. Fall applications for spring crabgrass control can be made after soil temperatures fall below 50°F but before the ground freezes.
Crew (150–200 lb) NY	dithiopyr + isoxaben (see label)	annual grasses and annual broadleaf weeds in turf	established bermudagrass, buffalograss, creeping bentgrass, fine fescue, Kentucky bluegrass, perennial ryegrass, tall fescue, zoysiagrass	Controls a wide range of annual grasses and annual broadleaf weeds. Has preemergence and early postemergence activity on crabgrass up to 1 tiller. See label for list of bentgrass and fine fescue cultivars prone to injury. Do not apply to putting greens or to newly seeded turf until it is well established. Irrigation or rainfall within 7 days after application will improve the performance of Crew. Do not reseed within 12 weeks of application.

W This product is labeled for use primarily in warm-season turf. Read the label for a list of tolerant grasses.
NY Use or sale of this product has restrictions in New York State or Long Island, New York (Nassau and Suffolk counties). Refer to the label for more information.
L This product is registered only in specific North Central states. Check with local contacts for availability.

Preemergence Heribicides (*continued*)

Trade Name (product/A)	Common Name (lb ai/A)	Weeds Controlled	Turfgrasses	Comments
Defendor SC (4 fl oz) NY	florasulam (0.013)	broadleaves such as catchweed bedstraw, common chickweed, dandelion, prickly lettuce, white clover, wild carrot, others	established bermudagrass, creeping bentgrass, fine fescue, Kentucky bluegrass, perennial ryegrass, tall fescue, zoysiagrass	Defendor has excellent soil and foliar activity at cool temperatures and provides fall and early spring postemergence broadleaf weed control. It is often mixed with a preemergence herbicide such as Dimension 2EW. This herbicide also prevents flowering in some broadleaves (including dandelion) when applied at least 1 week before flowering in spring.
Dimension 2EW (1.0–2.0 pt) Dimension Ultra 40 WP (0.625–1.25 lb) Dithiopyr 40 WSB (0.625–1.25 lb) Dimension and dithiopyr G available on fertilizer carrier (see label) others NY	dithiopyr (0.25–0.5)	summer annual grasses, annual bluegrass, yellow woodsorrel, some small-seeded broadleaves	established bermudagrass, buffalograss, creeping bentgrass, fine fescue, Kentucky bluegrass, perennial ryegrass, tall fescue, zoysiagrass	Has preemergence and early postemergence activity on crabgrass up to 1 tiller. Do not reseed, overseed, or sprig within 3 months of application. See label for specific programs related to mowing height, timing, and split applications. See label for list of bentgrass and fine fescue cultivars prone to injury.
Echelon 4SC (8–36 fl oz) Echelon G available on fertilizer carrier (see label) NY	sulfentrazone + prodiamine (0.25–1.125)	summer annual grasses, annual bluegrass, sedges, henbit, common chickweed, spurge, and some small-seeded broadleaves	established bermudagrass, buffalograss, creeping bentgrass, fine fescue, Kentucky bluegrass, perennial ryegrass, tall fescue, zoysiagrass	Apply only to well-established turf. Do not apply to creeping bentgrass mown < 0.5 inch. The use rate varies by turf species, so consult label for appropriate use rate. Wait 3–6 months before reseeding following an application. Do not use on sands with < 1% organic matter.
FreeHand 1.75G (100–200 lb) W NY	dimethenamid + pendimethalin (see label)	summer annual grasses, annual bluegrass, sedges, small-seeded broadleaves	established bermudagrass, buffalograss, zoysiagrass	**Do not use on cool-season turf.** Apply to well-established turf. Delay seeding at least 3 months following the last application. Apply to warm-season turfgrass following spring transition when soil temperature is 55°F or warmer. Rainfall or irrigation (0.5 inch) after application will improve control.
Gallery 75 DF (0.66–1.33 lb) Isoxaben 75WG (0.66–1.33 lb) Gallery SC (16 to 31 fl oz) NY	isoxaben (0.5–1.0)	broadleaves such as chickweed, henbit, spurge, plantain, others	established bermudagrass, buffalograss, creeping bentgrass, fine fescue, Kentucky bluegrass, perennial ryegrass, tall fescue, zoysiagrass	Controls a wide range of annual broadleaf weeds. Tank-mix with a grass herbicide to broaden spectrum of control. Do not apply to putting greens or to newly seeded turf until it is well rooted. Do not reseed within 60 days of application.
Gemini G (100–200 lb) Gemini 3.7SC (18–87 fl oz) NY	prodiamine + isoxaben (see label)	annual grasses and annual broadleaf weeds in turf	established bermudagrass, buffalograss, creeping bentgrass, fine fescues, Kentucky bluegrass, perennial ryegrass, tall fescue, zoysiagrass	Controls a wide range of annual grasses and annual broadleaf weeds. Do not apply to golf course putting greens or creeping bentgrass <0.5 inch tall. Rainfall or irrigation (0.5 inch) after application will improve control. Check the Gemini 3.7SC label for rate recommendations on individual turf species. Wait 60 days after seeding or until after the second mowing, whichever is longer, before applying Gemini. Wait 4–6 months before reseeding following an application.
Goosegrass/ Crabgrass Control (114 lb)	bensulide + oxadiazon (6.0 + 1.5)	annual grasses	established bermudagrass, creeping bentgrass, Kentucky bluegrass, perennial ryegrass, tall fescue, zoysiagrass	Apply a light irrigation after treatment. See label for precautions about use on putting greens. Limit of 2 applications per year on putting greens. Do not reseed for 5 months after application.
Kerb 50 WP (1–4 lb) Kerb SC T&O (1.25–5.0 pt) W	pronamide (0.5–2.0)	annual bluegrass, perennial ryegrass, as well as other grassy and broadleaf weeds in warm-season turf	established bermudagrass, buffalograss, zoysiagrass	**Do not use on cool-season turf.** Restricted use pesticide. Used for preemergence and postemergence control in established turf. Works slowly (3–5 weeks). Do not apply to cool-season grasses or slopes that may drain onto cool-season grasses. A light rain or irrigation is needed to activate if no rain falls 24–48 hours of application. Limit of 1 application per season.

W This product is labeled for use primarily in warm-season turf. Read the label for a list of tolerant grasses.
NY Use or sale of this product has restrictions in New York State or Long Island, New York (Nassau and Suffolk counties). Refer to the label for more information.

Preemergence Heribicides (*continued*)

Trade Name (product/A)	Common Name (lb ai/A)	Weeds Controlled	Turfgrasses	Comments
Pendulum 2G, (see label) Pendulum Aquacap (see label) Pendulum 3.3 EC (see label) pendimethalin G available on fertilizer carrier (see label) others	pendimethalin (1.5–3.0)	summer annual grasses, annual bluegrass, yellow woodsorrel, some small-seeded broadleaves	established annual bluegrass, bermudagrass, buffalograss, creeping bentgrass, fine fescue, Kentucky bluegrass, perennial ryegrass, tall fescue, zoysiagrass	Not recommended for turfgrass that has been severely thinned due to winter stress. Do not reseed within 4 months of application. See label for specific rates for certain weeds and turf species. Sequential applications can be made for extended control. See Pendulum Aquacap and Pendulum 2G labels for instructions on 0.5 inch or taller annual bluegrass and creeping bentgrass.
Pennant MAGNUM (1.3–2.6 pt) W NY	metolachlor (1.24–2.48)	annual bluegrass, crabgrass, and sedges	established bermudagrass, zoysiagrass	**Do not use on cool-season turf.** Apply before weeds emerge and irrigate in with 0.5 inch of water if rainfall does not occur within 7 days. Delayed spring green-up, temporary slowing of growth, and yellowing may occur following application. Do not apply to newly seeded areas. Do not reseed within 4 months of application.
Princep Liquid (1–1.6 qt) Simazine 90DF (1.1–1.8 lb) others W NY	simazine (1.0–1.6)	summer annual grasses, annual bluegrass, henbit, common chickweed, and some small-seeded broadleaves in warm-season turf	established bermudagrass, buffalograss, zoysiagrass	**Do not use on cool-season turf.** Apply only to dormant turf. Apply between Nov. 1 and Dec. 15 for winter weed control. Do not apply to areas that drain onto golf course tees, putting greens, or other desirable turfgrasses. Avoid application during spring green-up. Do not exceed 3 qt/A per calendar year. Limit of 2 applications per year. Irrigate immediately following application to turf.
PoaCure (0.2–0.4 gal) NY	methiozolin (0.46–0.92)	Preemergence and postemergence control of annual bluegrass. Preemergence crabgrass and goosegrass control.	bermudagrass, creeping bentgrass, fine fescue, Kentucky bluegrass, perennial ryegrass, tall fescue, zoysiagrass	**For use on golf courses only.** Water in sprays with 0.1 to 0.2-inch of irrigation after application. The herbicide is slow acting, taking 2 to 4 weeks to control weeds. Do not treat putting greens under stress or within one week of aerification. Check the label for application timing, rate, and frequency recommendations.
Prograss (0.5–1.33 gal) Prograss SC, (0.75–2 qt)	ethofumesate (0.75–2.0)	annual bluegrass, annual grasses, and some annual broadleaves	creeping bentgrass, dormant bermudagrass, perennial ryegrass, Kentucky bluegrass, tall fescue	See label for recommendations for reducing annual bluegrass in cool-season turf. Do not use more than 0.75 lb ai/A on Kentucky bluegrass or creeping bentgrass. Refer to the label for further instructions on each grass species.
Ronstar 50 WP (4–8 lb) Ronstar Flo 3.17L (2.5–3.8 qt) Ronstar and oxadiazon G available on fertilizer carrier (see label) others	oxadiazon (see label)	summer annual grasses including goosegrass, annual bluegrass, some small-seeded broadleaves	established bermudagrass, buffalograss, creeping bentgrass, fine fescue, Kentucky bluegrass, perennial ryegrass, tall fescue, zoysiagrass	Granular formulations are the only formulations labeled for use on cool-season grasses. Ronstar G is no longer available, but many other granular formulations are available. Apply to dormant bermudagrass or zoysiagrass at least 2–3 weeks before green-up. Do not apply to wet turf. Irrigate immediately after application. Delay reseeding until 4 months after treatment. Not for use in residential turf. An EPA issued Proposed Interim Decision is set to designate oxadiazon as a Restricted Use Pesticide and allow treatment of no more than 30% of all turf surface on a particular property. There will be other less impactful restrictions as well. This will take effect no earlier than 2025.
Sencor 75% (0.33–0.66 lb) W L	metribuzin (0.25 to 0.5)	annual bluegrass, goosegrass, broadleaf annual weeds	bermudagrass	**Do not use on cool-season turf.** Use low rate on actively growing bermudagrass. Use high rate on dormant bermudagrass for winter annual weed control. Do not apply in the root zone of shallow-rooted ornamentals. Do not apply to golf course tees, putting greens or other turf mowed at < ½ inch.
Specticle FLO (3–10 fl oz) Specticle G (100–200 lb/A) W NY	indaziflam (see label)	annual grassy and broadleaf weeds in turf	established bermudagrass, buffalograss, zoysiagrass	**Do not use on cool-season turf unless thinning or removal is desired.** Do not apply to golf course putting greens, tees, or collars. Irrigate immediately after application. Use a 15-foot buffer area around cool-season grasses.

W This product is labeled for use primarily in warm-season turf. Read the label for a list of tolerant grasses.
NY Use or sale of this product has restrictions in New York state or Long Island, New York (Nassau and Suffolk counties). Refer to the label for more information.
L This product is only registered in specific North Central states. Check with local contacts for availability.

Preemergence Heribicides (*continued*)

Trade Name (product/A)	Common Name (lb ai/A)	Weeds Controlled	Turfgrasses	Comments
SureGuard WDG (8–12 oz) SureGuard SC (8–12 fl oz) StayGuard on fertilizer carrier (see label) W	flumioxazin (0.25–0.38)	annual bluegrass, winter annual broadleaf weeds, summer annual grasses	dormant bermudagrass, zoysiagrass (SureGuard SC formulation only)	**Do not use on cool-season turf.** Apply only to dormant bermudagrass just after winter dormancy begins to control annual bluegrass or in late winter well before green-up for summer annual grassy weed control. Apply at least 15 feet upslope from creeping bentgrass putting greens.
Surflan AS (1.5–2.0 qt) Surflan WDG (1.75–2.4 lb) Oryzalin 4 (1.5–2.0 qt) others W	oryzalin (1.5–2.0)	summer annual grasses, annual bluegrass, some small-seeded broadleaves	established bermudagrass, buffalograss, tall fescue, zoysiagrass	Do not use on cool-season turf except tall fescue. Apply for summer annual grass control. Split applications are recommended for improved goosegrass control and for tall fescue. Do not apply to tall fescue < 6 months old. Do not apply to newly sprigged grasses until well established. Do not make a spring application to fall-planted turfgrasses. Delay reseeding 90–120 days after application.
Surflan XL 2G (100–150 lb) W	benefin + oryzalin (1.0 + 1.0–1.5 + 1.5)	summer annual grasses, annual bluegrass, some small-seeded broadleaves	established bermudagrass, buffalograss, tall fescue, zoysiagrass	Do not use on cool-season turf except tall fescue. Apply for summer annual grass control. Split applications are recommended for improved goosegrass control and for tall fescue. Do not apply to tall fescue < 6 months old. Do not apply to newly sprigged grasses until well established. Do not make a spring application to fall-planted turfgrasses. Delay reseeding at least 6–16 weeks after application.
Team Pro (75–150 lb) Crabgrass Control 2% Team (100 lb) Team also available on fertilizer carrier (see label)	benefin + trifluralin (see label)	summer annual grasses, annual bluegrass, some small-seeded broadleaves	established bermudagrass, buffalograss, creeping bentgrass, fine fescue, Kentucky bluegrass, perennial ryegrass, tall fescue, zoysiagrass	Do not apply to newly sprigged grasses until well established. Do not use on putting greens or bentgrass mown < 0.5 inch. Delay reseeding for 8 weeks (if applied at a low rate) and for 12–16 weeks (high rate) after application.
Tower (21–32 fl oz) W NY	dimethenamid (1.0–1.5)	bittercress, crabgrass, goosegrass, purslane, sedges, spurge	established bermudagrass, buffalograss, zoysiagrass	On golf courses: can be used on cool- and warm-season turf. **For all other turf areas: warm-season turf only.** Apply to well-established turf or new seedlings that have been mown at least 4 times. Delay seeding at least 1 month after last application. Do not apply more frequently than every 35 days and do not apply more than 64 fl oz (3.0 lb ai) per year.

W This product is labeled for use primarily in warm-season turf. Read the label for a list of tolerant grasses.
NY Use or sale of this product has restrictions in New York State or Long Island, New York (Nassau and Suffolk counties). Refer to the label for more information.

Postemergence Herbicides

Weed Control Ratings for Postemergence Broadleaf Herbicides

Herbicide	Canada thistle	chickweed, common	chickweed, mouse-ear	clover, white	dandelion	garlic, wild (onion)	henbit	ground ivy	knotweed, prostrate	lespedeza	mallow	plantains	purple deadnettle	purslane, common	shepherd's purse	speedwell, corn	spurge, prostrate	strawberry, Indian mock	wild violet	yellow woodsorrel
2,4-D (Weedar64, Weedone LV4, others)	G	P	P	P	E	G	P	F	P	P	G	G	G	G	G	P	P	F	P	P
2,4-D + aminopyralid (NativeKlean)	G	P	P	P	E	G	P	F	P	P	G	G	G	G	G	P	P	F	P	P
2,4-D + dicamba (On Deck)	G	P	E	G	E	G	E	F	G	P	G	G	G	E	G	P	G	F	P	P
2,4-D + dichlorprop (2,4-DP)	G	G	G	F	E	G	E	F	F	G	G	G	G	G	G	F	G		P	P
2,4-D + clopyralid + dicamba (Millennium Ultra 2)	E	G	E	E	E	G	E	F	G	G	G	E	G	G	G	P	G	F	P	P
2,4-D + dichlorprop + dicamba (Brushmaster)	G	G	E	G	E	G	E	F	G	G	E	G	G	G	G		F		P	P
2,4-D + fluroxypyr + dicamba (Escalade 2)	G	G	E	G	E	G	E	G	G	E	G	G	G	E	G	P	G	F	P	E
2,4-D + fluroxypyr + halauxifen-methyl (GameOn)	E	G	E	G	E	G	E	G	G	E	G	E	G	E	G	P	G	F	P	E
2,4-D + fluroxypyr + triclopyr + sulfentrazone (Momentum 4-Score)	G	E	E	G	E	G	E	G	G	E	G	G	G	E	G	G	F		G	G
2,4-D + MCPA + dicamba (Triad Select)	G	E	E	E	E	G	E	F	G	G	E	G	G	G	E	F	F	E	P	P
2,4-D + MCPA + dicamba + sulfentrazone (Triad SFZ Select)	G	E	E	E	E	G	E	F	G	G	E	G	G	G	E	F	F	E	P	P
2,4-D + MCPA + triclopyr + dicamba (Triad T Select)	G	G	E	G	E	G	E	G	G	E	G	G	G	E	G	G	G	F	G	G
2,4-D + MCPP + dichlorprop (Triamine)	G	E	E	E	E	G	E	F	F	G	E	G	F	G	E	F	F		P	P
2,4-D + MCPP + dicamba (Trimec 992, Triplet, others)	G	E	E	E	E	G	E	F	G	G	E	G	G	G	E	F	F	E	P	P
2,4-D + MCPP + dicamba (Trimec Southern)	G	E	E	E	F	F	E	P	G	G	E	F	G	G	E	F	F	G	P	P
2,4-D + MCPP + dicamba + pyraflufen-ethyl (RedZone 2)	G	E	E	E	E	G	E	F	G	G	E	G	G	G	G	F	G	E	P	P
2,4-D + quinclorac + dicamba (2DQ, Momentum Q, Quincept)	G	F	E	E	E	G	E	F	G	G	G	G	G	G	G	P	G	F	P	P
2,4-D + triclopyr + dicamba + pyraflufen-ethyl (4-Speed XT)	E	E	E	E	E	G	E	G	G	E	E	G	G	E	G	G	G	E	P	G
2,4-D + triclopyr (Chaser, others)	G	E	E	G	E	G	E	G	G	E	G	G	F	E	G	G	F		G	E
2,4-D + triclopyr + dicamba + sulfentrazone (TZONE SE)	G	E	E	E	E	G	E	G	G	E	G	G	E	E	E	E	G	G	G	E
2,4-D + triclopyr + fluroxypyr (Momentum FX2)	G	E	E	G	E	G	E	G	F	E	G	G	G	E	G	G	F		G	G
2,4-D + triclopyr + fluroxypyr + flumioxazin (Sure Power)	G	E	E	G	E	G	E	G	F	E	G	G	G	E	G	G	F		G	G
amicarbazone (Xonerate 2SC)		G					G							G	G	G				
atrazine (AAtrex)	P	E	E	G	P	P	E			F		P	E	G		G	G			
bentazon (Basagran)	G						P				P		P	G	G					P
bromoxynil (Buctril)	G					P	G	P	F		F	P			G	G	G		P	F
carfentrazone (Quicksilver T&O)	F	F	F	F	F		F	P			G	P		G	G	G	F			

Weed Control Ratings for Postemergence Broadleaf Herbicides (*continued*)

Herbicide	Canada thistle	chickweed, common	chickweed, mouse-ear	clover, white	dandelion	garlic, wild (onion)	henbit	ground ivy	knotweed, prostrate	lespedeza	mallow	plantains	purple deadnettle	purslane, common	shepherd's purse	speedwell, corn	spurge, prostrate	strawberry, Indian mock	wild violet	yellow woodsorrel
carfentrazone + 2,4-D + MCPP + dicamba (SpeedZone)	G	E	E	E	E	G	E	F	F	G	G	G	G	G	G	F	F	E	P	P
carfentrazone + MCPA + MCPP + dicamba (PowerZone)	G	E	E	E	G	G	E	F	F	G	G	G	G	G	G	F	F	E	P	P
chlorsulfuron (Telar XP)	G	G	G	G		G	E		G		G	F		F	G		G		G	
clopyralid (Lontrel)	E	P	P	E	F			P	G	P		E		F	F	G	P	P	P	P
dicamba (Banvel, Vanquish, Diablo)	G	E	E	E	P	G	G	P	G		G	P	G	E	G	P	G		P	P
dicamba + metsulfuron (Fahrenheit)	G	E	E	E	E	E	E	F	G	E	G	G	G	E	G	E	E		G	E
ethofumesate (Prograss)		F		F										F						
fenoxaprop + fluroxypyr + dicamba (Last Call)	G	G	G	E	G	G	G	F	G	G				G					P	
flazasulfuron (Katana)		E	G	G	G	P	E			P			G	G					G	E
florasulam (Defendor SC)	F	E		G	G	P	P	P	F			P	P		E		P		P	
fluroxypyr (Vista XRT)		E	G	E	F		E	G	F	E		F	G	E			G		P	E
fluroxypyr + triclopyr (Tailspin)	G	G	G	G	G		E	G	G	E		G	E	E		G	G		G	E
glyphosate	F	E		P	P		F						F	E	E				P	
halosulfuron + dicamba (Yukon[1])	G		E	G	P	G	E	P	G		G	P	G	G	G		G			
imazapic (Plateau)		F			F	F	F		F			F		F			F			F
imazaquin (Image 70DG)		G	G	G	P	G	G						G						P	
imazosulfuron (Celero)		G	G				G							G						
MCPA	G	E		F	G				G		F	F		G					P	F
MCPA + fluroxypyr + dicamba (Change Up)	G	G	G	E	G	G	G	F	G	E	F	G	G	E	G	G	G		P	G
MCPA + fluroxypyr + dichlorprop (Chaser Ultra 2)	G	G	G	G	G	G	G	F	G	E	F	G	G	E	G	G	G		P	G
MCPA + fluroxypyr + triclopyr (Battleship III)	G	G	G	G	G	E	E	G	G	E	F	G	E	E	G	G	G		G	E
MCPA + MCPP + dicamba (Trimec Encore)	G	G	G	G	G	G	G	F	G	G	G	G	G	G	E	G	G	G		P
MCPA + triclopyr + dicamba (Cool Power, Horsepower, others)	G	G	E	G	G	G	E	G	G	E	G	F	G	E	G	G	G		G	G
MCPP, mecoprop (Mecomec)	G	F	F	F	P	P	F	P	P		F	F	F	F	E	P	P		P	P
mesotrione (Tenacity)	F	F	F	F	F		F	P				F		P		F			F	F
metribuzin (Sencor 75%)		E	E	G			E		G	F			E		G	E	G			
metsulfuron (Manor, Mansion, MSM)	P	E	E	E	E	E	E	G	E	E		G	P	E	G	E	E		G	E
metsulfuron + rimsulfuron (Negate 37WG)	P	E	E	E	E	E	E	G	E	E		G	P	E	G	E	E		G	E

Weed Control Ratings for Postemergence Broadleaf Herbicides (*continued*)

Herbicide	Canada thistle	chickweed, common	chickweed, mouse-ear	clover, white	dandelion	garlic, wild (onion)	henbit	ground ivy	knotweed, prostrate	lespedeza	mallow	plantains	purple deadnettle	purslane, common	shepherd's purse	speedwell, corn	spurge, prostrate	strawberry, Indian mock	wild violet	yellow woodsorrel
penoxsulam (LockUp, Sapphire)		E	G	G	G			P		G		P		P	G					
penoxsulam + 2,4-D + dicamba (LockUp Extra 2)	F	E	G	G	G			P	G	G		G		P	G	F				
penoxsulam + sulfentrazone + 2,4-D + dicamba (Avenue South)	F	E	G	E	E	G	G	F	G	G	G	F	G	G	G	F	G	F		F
prodiamine + quinclorac (Cavalcade PQ)				E	E			F		P		G		P					P	
pyraflufen ethyl (Octane)		G	G	G	G		G		G		G			P	G	G				
pyrimisulfan (Arkon, Vexis)		G				G	G	F				F	G							
quinclorac (Drive XLR8)		P		E	F			F		P		G		P					P	P
quinclorac + sulfentrazone + 2,4-D + dicamba (Q4 Plus)	G	G	E	E	E	G	E	F	G	G	G	G	E	G	E	G	G	G	F	P
rimsulfuron (Rimsulfuron 25DF)		G					E						G				F			P
simazine (Princep)	P	E	E	G			E			F			E	F			F			
sulfentrazone (Dismiss, Spartan 4F[1])		G		P	P		G	P	P	P	G	P		F		G	G	G	F	P
sulfentrazone + carfentrazone (Dismiss NXT)	F	G	F	F	F		G	P	F		G	G		F	G	G	G	G	P	P
sulfentrazone + metsulfuron (Blindside)	P	E	E	E	E	E	E	G	E	E	G	G		E	G	E	E	G	G	P
sulfentrazone + quinclorac (Solitare)		G		E	E		G	F	G	P	G			F		G	G	G	F	P
sulfentrazone + 2,4-D + MCPP + dicamba (Surge)	G	G	E	E	E	G	E	F	G	G	G	G	G	G	E	G	G	G	F	P
sulfosulfuron (Certainty)		G		F	F	E	P	F						P	G		P			E
thiencarbazone + foramsulfuron + halosulfuron (Tribute TOTAL)		G	G	G			G					G			G	G	F			F
thiencarbazone + iodosulfuron + dicamba (Celsius WG)	G	E	G	E	E	G	E	G	G	G		F		G	G	G	G			G
thiencarbazone + iodosulfuron + halosulfuron (Celsius XTRA)	P	G	G	G	F	G	G	F	G	F		G	G	F	G	G	P		F	F
topramezone (Pylex)	G	G		G	G			G			G	F		P	G					
triclopyr (Turflon Ester Ultra)	G	P	F	F	P			G	G	E		F		E		G	G	G	G	E
triclopyr + clopyralid (Confront, 2-D)	E	P	F	E	F	F	F	G	G	E	F	E	F	E	F	E	G	G	G	E
triclopyr + dicamba + mesotrione (Sublime)	G	E	E	E	F	G	G	G	G	E	G	F	G	E	G	G	G	G	G	E
trifloxysulfuron (Monument, Recognition)				G	G		G					P					E		P	E

Rating Key: E = excellent (≥90% control). G = good (75%–90% control). F = fair (50%–75% control). P = poor control (≤50% control). Blank = no data.

1 Sod farm use only.

Weed Control Ratings for Postemergence Grass Herbicides

Herbicide	annual bluegrass	barnyardgrass	bermudagrass	crabgrass	creeping bentgrass	dallisgrass	field paspalum	fountain grass	goosegrass	Japanese stiltgrass	johnsongrass	nimblewill	quackgrass	rough bluegrass	ryegrasses	sandbur	tall fescue	yellow foxtail	windmillgrass	zoysiagrass
2,4-D + MCPP + dicamba + carfentrazone (SpeedZone)				P					G			G				P				
2,4-D + quinclorac + dicamba (Momentum Q, Quincept)	P	G	P	G		P	P	G	P	P					P	P	P	G		
amicarbazone (Xonerate 2SC)	F			P		P				E				P						
asulam (Asulox)				G					G							G				
atrazine (AAtrex)	E			P		P	P		P						G	F	F			
chlorsulfuron (Telar XP)	F				P										E		G			P
ethofumesate (Prograss)	G		P	P		P	P		P						P	P	P			P
fenoxaprop (Acclaim Extra)	P		F	G		P	P	G	G	E	G				P	E	P	G	F	
fenoxaprop + fluroxypyr + dicamba (Last Call)	P		F	G		P	P	G	G	E	G				P	E	P	G	F	
flazasulfuron (Katana)	G	G		F	G	P	P		P				E	E	E		E		P	
fluazifop (Fusilade II, Ornamec)	P	G	F	G		G	G	P	G	G	G		F		F	G		G		
foramsulfuron (Revolver)	E		P		F	F	F		G					G	E	P	G		P	
glyphosate[1]	E	E	F	E	F	F	F	E	E	E	G	F	F	F	E	E	E	E	F	G
glyphosate + 2,4-D (Campaign)	E	E	F	E	F	F	F	G	E	E	G	F	F	F	E	E	E	E	F	G
imazapic (Plateau)				G		F	F		G	P	F				F	G	F	G		
imazaquin (Image 70DG)	F			F											G	F	F			
mesotrione (Tenacity)	F		P	G	G	P	F		F	F		G				P		P	F	P
methiozolin (PoaCure)	E													G						
metribuzin (Sencor 75%)	G			F		F	F		G						G	F				
metsulfuron (Manor, Mansion)	F		P	F	P	P	P		P	P	F				E	P	F	G		
metsulfuron + rimsulfuron (Negate 37WG)	G		P	F		P	P		P		F			E	G	P	P	G		
MSMA	P	G	P	G		F	F		F	E	F				P	E	P	F		
pinoxaden (Manuscript)				G		G	G			P										
prodiamine + quinclorac (Cavalcade PQ)	P	G	P	G		P	P	G	P						P	P	P	G	P	
pronamide (Kerb)	G	G	P	P	G	P	P		P					G	G	P	P	G		
quinclorac (Drive XLR8)	P	G	P	G		P	P	G	P	P					P	P	P	G	P	
quinclorac + sulfentrazone + 2,4-D + dicamba (Q4 Plus)	P	G	P	G		P	P	G	P	P					P	P	P	G		
rimsulfuron (Rimsulfuron 25DF)	E		P	P	G	P	P		P					E	E		E			
sethoxydim (Segment II)	P	G	G	G	F	P	P	P	G	E	F		G	E	P	G	P	E		

Weed Control Ratings for Postemergence Grass Herbicides (*continued*)

Herbicide	annual bluegrass	barnyardgrass	bermudagrass	crabgrass	creeping bentgrass	dallisgrass	field paspalum	fountain grass	goosegrass	Japanese stiltgrass	johnsongrass	nimblewill	quackgrass	rough bluegrass	ryegrasses	sandbur	tall fescue	yellow foxtail	windmillgrass	zoysiagrass
simazine (Princep)	G		P	F		P	P		P						G	F	P			
simazine + prodiamine + imazaquin (Coastal)	G		P	F		P	P		P						G	F	F			P
sulfentrazone (Dismiss)				P					F							P			P	
sulfentrazone + imazethapyr (Dismiss South)				G					G							G				
sulfentrazone + quinclorac (Solitare)		G		G		P	P	G	F	P						P		G	P	
sulfentrazone + 2,4-D + MCPP + dicamba (Surge)				P					F											
sulfosulfuron (Certainty)	G		P	P	P	P	P		P		G		F	G	G		G			
thiencarbazone + foramsulfuron + halosulfuron (Tribute TOTAL)	E			F	E	G	F		F					E	E		G			
thiencarbazone + iodosulfuron + dicamba (Celsius WG)		E				P	P			P	F		E		F			G	P	
thiencarbazone + iodosulfuron + halosulfuron (Celsius XTRA)										F					F			F		
topramezone (Pylex)	P	G	G	G	P	F	F		E	G	P	G				E		E	E	F
triclopyr (Turflon Ester Ultra)			P	P	P					P									G	
triclopyr + sulfentrazone + 2,4-D + dicamba (TZONE SE)			P						F	P									G	
trifloxysulfuron (Monument, Recognition)	E			P	F	P	F		P						E	P	G		P	

Rating Key: E = excellent (≥90% control). G = good (75%–90% control). F = fair (50%–75% control). P = poor control (≤50% control). Blank = no data.

1 Use only in completely dormant bermudagrass or as a preplant application.

Turfgrass Tolerance to Postemergence Herbicides

Herbicide	Cool-Season						Warm-Season		
	annual bluegrass	creeping bentgrass	fine fescue	Kentucky bluegrass	perennial ryegrass	tall fescue	bermudagrass	buffalograss	zoysiagrass
2,4-D (Weedar64, Weedone LV4, others)	S	S[1]	S	S	S	S	S	S-I	S
2,4-D + aminopyralid (NativeKlean)	S	S	S	S	S	S	S-I	S-I	S
2,4-D + dicamba (On Deck)	S	S[1]	NR	S	S	S	S	S-I	S
2,4-D + dichlorprop (Patron 170)	S	S-I	S	S	S	S	S	S-I	S
2,4-D + clopyralid + dicamba (Millennium Ultra 2)	S	S	S	S	S	S	S	S-I	S
2,4-D + dichlorprop + dicamba (Brushmaster)	S	NR	S	S	S	S	S-I	S-I	S-I
2,4-D + fluroxypyr + dicamba (Escalade 2)	S	S	S	S	S	S	S	NR	S
2,4-D + fluroxypyr + halauxifen-methyl (GameOn)	NR	S	S	S	S	S	S	S-I	S
2,4-D + triclopyr + fluroxypyr (Momentum FX2)	NR	S	S	S	S	S	S	I	S
2,4-D + MCPA + dicamba (Triad Select)	S	S	S	S	S	S	S	NR	S
2,4-D + MCPA + dicamba + sulfentrazone (Triad SFZ Select)	S	S	S	S	S	S	S	S-I	S
2,4-D + MCPA + triclopyr + dicamba (Triad T Select)	S	NR	S	S	S	S	S-I	NR	S-I
2,4-D + MCPP + dichlorprop (Triamine)	S	S	S	S	S	S	S	S-I	S
2,4-D + MCPP + dicamba (Trimec, others)	S	S	S	S	S	S	S	S-I	S
2,4-D + MCPP + dicamba + pyraflufen-ethyl (RedZone 2)	S	S	S	S	S	S	S	S-I	S
2,4-D + quinclorac + dicamba (2DQ)	NR	S	NR	S	S	S	S	S-I	S
2,4-D + quinclorac + dicamba (Momentum Q, Quincept)	S	S-I	S-I	S	S	S	S	NR	S
2,4-D + fluroxypyr + triclopyr + sulfentrazone (Momentum 4-Score)	S	S	S	S	S	S	NR	NR	NR
2,4-D + triclopyr + dicamba + pyraflufen-ethyl (4-Speed XT)	S	S	S	S	S	S	S	NR	S
2,4-D + triclopyr (Chaser, Chaser 2 Amine, Turflon II Amine)	S	S-I	S	S	S	S	NR	NR	NR
2,4-D + triclopyr + dicamba + sulfentrazone (Foundation)	S	NR	S	S	S	S	NR	NR	NR
2,4-D + triclopyr + fluroxypyr + flumioxazin (Sure Power)	NR	NR	S	S	S	S	NR	NR	NR
amicarbazone (Xonerate 2SC)	NR	S	S	S	S	S	S	S	S
asulam (Asulox[2])	NR	NR	NR	NR	NR	NR	S-I	NR	NR
atrazine (AAtrex)	NR	NR	NR	NR	NR	NR	S-I	NR	S-I
bentazon (Basagran, LescoGran)	S	S	S	S	S	S	S	S	S
bromoxynil (Buctril)	NR	S	S	S	S	S	S	NR	S
carfentrazone (Quicksilver T&O)	S	S	S	S	S	S	S	S	S
carfentrazone + 2,4-D + MCPP + dicamba (SpeedZone)	S	S-I	S	S	S	S	S	S	S
carfentrazone + MCPA + MCPP + dicamba (PowerZone)	S	NR	S	S	S	S	S-I	NR	S
chlorsulfuron (Telar XP)	NR	S	S	S	NR	NR	S	S	NR
clopyralid (Lontrel)	NR	S	S	S	S	S	S	S	S
dicamba (Banvel, Vanquish, Diablo)	S	S-I	S	S	S	S	S	S-I	S
dicamba + metsulfuron (Fahrenheit)	NR	NR	NR	NR	NR	NR	S	S	S
ethofumesate (Prograss)	NR	S[2]	NR	S[2]	S	S	S[3]	NR	NR
fenoxaprop (Acclaim Extra)	S	S-I	S	S	S	S	NR	NR	S
fenoxaprop + fluroxypyr + dicamba (Last Call)	NR	NR	S	S	S	S	NR	NR	S
flazasulfuron (Katana)	NR	NR	NR	NR	NR	NR	S	S	S
florasulam (Defendor SC)	S	S	S	S	S	S	S	NR	S
fluazifop-p (Fusilade II, Ornamec)	S	NR	S	NR	NR	S-I	NR	NR	S-I
fluroxypyr (Vista XRT)	NR	S-I	S	S	S	S	S-I	NR	S
fluroxypyr + triclopyr (Tailspin)	NR	S-I	S	S	S	S	NR	NR	S
foramsulfuron (Revolver)	NR	NR	NR	NR	NR	NR	S	S	S
glyphosate + 2,4-D (Campaign)	NR	NR	NR	NR	NR	NR	S[3]	NR	NR
halosulfuron + dicamba (Yukon[2])	S	S	S	S	S	S	S	S-I	S
imazapic (Plateau)	NR	NR	NR	NR	NR	NR	S	S	NR
imazaquin (Image 70DG)	NR	NR	NR	NR	NR	NR	S	NR	S
imazosulfuron (Celero)	NR	S	S	S	S	S	S	NR	S
MCPA	S	S	S	S	S	S	S	S-I	S
MCPA + fluroxypyr + dicamba (Change Up)	S	S	S	S	S	S	S	NR	S

Turfgrass Tolerance to Postemergence Herbicides (*continued*)

Herbicide	Cool-Season						Warm-Season		
	annual bluegrass	creeping bentgrass	fine fescue	Kentucky bluegrass	perennial ryegrass	tall fescue	bermudagrass	buffalograss	zoysiagrass
MCPA + fluroxypyr + dichlorprop (Chaser Ultra 2)	S	S	S	S	S	S	S	NR	S
MCPA + fluroxypyr + triclopyr (Battleship III)	NS	S-I	S	S	S	S	S-I	S-I	S
MCPA + MCPP + dicamba (Trimec Encore)	S	S	S	S	S	S	S	S-I	S
MCPA + triclopyr + dicamba (Cool Power, Horsepower, others)	S	S	S	S	S	S	S-I	NR	S
MCPP, mecoprop (Mecomec)	S	S-I	S	S	S	S	S	NR	S
mesotrione (Tenacity)	NR	NR	S	S	S	S	I[2]	S	NR
methiozolin (PoaCure)	NR	S	S	S	S	S	S	NR	S
metribuzin (Sencor 75%)	NR	NR	NR	NR	NR	NR	S	NR	NR
metsulfuron (Manor, Mansion)	NR	NR	NR	S[4]	NR	NR	S	S	S
metsulfuron + rimsulfuron (Negate 37WG)	NR	NR	NR	NR	NR	NR	S	NR	S
MSMA	S-I	NR	NR	S-I	S-I	S-I	S-I	I	S-I
pinoxaden (Manuscript)	NR	NR	NR	NR	NR	NR	S	NR	S
penoxsulam (LockUp)	S	NR	S	S	S-I	S-I	S	NR	S
penoxsulam + 2,4-D + dicamba (LockUp Extra 2)	S	NR	S	S	S-I	S-I	S	NR	S
penoxsulam + sulfentrazone + 2,4-D + dicamba (Avenue South)	S	NR	S	S	S-I	S-I	S	S	S
prodiamine + quinclorac (Cavalcade PQ)	NR	S-I	S-I	S	S	S	S	S	S
pronamide (Kerb)	NR	NR	NR	NR	NR	NR	S	S	S
pyraflufen ethyl (Octane)	NR	S	NR	S	S	S	S	S	S
pyrimisulfan (Arkon, Vexis)	NR	S	S	S	S	S	S	S	S
quinclorac (Drive, others)	S	S-I	S-I	S	S	S	S	S	S
quinclorac + sulfentrazone + 2,4-D + dicamba (Q4 Plus)	S	NR	S	S	S	S	S	NR	S
rimsulfuron (Rimsulfuron 25DF)	NR	NR	NR	NR	NR	NR	S	S	S
simazine (Princep)	NR	NR	NR	NR	NR	NR	S-I	S	S-I
simazine + prodiamine + imazaquin (Coastal)	NR	NR	NR	NR	NR	NR	S[3]	NR	S[3]
sethoxydim (Segment II)	S	NR	S	NR	NR	NR	NR	I	NR
sulfentrazone (Dismiss, Spartan 4F[2])	NR	S-I	S-I	S	S	S-I	S	S	S[2]
sulfentrazone + 2,4-D + MCPP + dicamba (Surge)	S	S	S	S	S	S	S	S-I	S
sulfentrazone + carfentrazone (Dismiss NXT)	NR	S-I	S-I	S	S	S-I	S	S	S[2]
sulfentrazone + imazethapyr (Dismiss South)	NR	NR	NR	NR	NR	NR	S	S	S[2]
sulfentrazone + metsulfuron (Blindside)	NR	NR	NR	S[4]	NR	S-I	S	S	S[2]
sulfentrazone + quinclorac (Solitare)	NR	NR	S-I	S	S	S	S	S	S[2]
sulfosulfuron (Certainty)	NR	NR[4]	NR	NR[4]	NR	NR	S	S	S
thiencarbazone + foramsulfuron + halosulfuron (Tribute TOTAL)	NR	NR	NR	NR	NR	NR	S	NR	S
thiencarbazone + iodosulfuron + dicamba (Celsius WG)	NR	NR	NR	NR	NR	NR	S	S	S
thiencarbazone + iodosulfuron + halosulfuron (Celsius XTRA)	NR	NR	NR	NR	NR	NR	S	S	S
topramezone (Pylex)	NR	NR	S	S	S	S	NR	NR	NR
triclopyr (Turflon Ester Ultra)	S	NR	S	S	S	S	NR	I	S
triclopyr + clopyralid (Confront or 2-D)	NR	S-I	S	S	S	S	S-I	S-I	S
triclopyr + dicamba + mesotrione (Sublime)	NR	NR	S	S	S	S	S[3]	NR	S-I
triclopyr + sulfentrazone + 2,4-D + dicamba (TZONE SE)	S	NR	S	S	S	S	S	NR	S
trifloxysulfuron (Monument, Recognition)	NR	NR	NR	NR	NR	NR	S	S-I	S

Rating Key: S = safe at labeled rates. I = intermediate safety or some injury may occur, use at reduced rates. NR = not registered for use on this turfgrass; do not use.

1 Spot treatments only.
2 Safety varies by cultivar.
3 Can be used safely during winter dormancy.
4 Not safe for use on Kentucky bluegrass + perennial ryegrass or Kentucky bluegrass + fine fescue mixtures, but is safe for use on 100% Kentucky bluegrass.
5 A label change in 2011 no longer allows for the use of Certainty turf herbicide on cool-season turf. These changes are effective on product packaged and shipped after May 2011. All cool-season turfgrass uses for Certainty have been removed from the 2011-2012 label (product packaged and shipped after May 2011). Product packaged before these changes can continue to be used according to label directions in cool-season turf. Previous label stipulations recommended that Certainty was safe for use on creeping bentgrass, 100% Kentucky bluegrass turf, or for use on Kentucky bluegrass + perennial ryegrass and/or fine fescue mixtures, but not safe for use on tall fescue + Kentucky bluegrass mixtures.

Postemergence Herbicides

Refer to herbicide labels for weeds controlled and specific turfgrass species use instructions.

Trade Name (product/A)	Common Name (lb ai or ae/A)	Weeds Controlled	Turfgrasses	Comments
Amine Formulations 2,4-D Amine 4 (2–4 pt) Weedar 64 (2–3 pt) others **Ester Formulations** Weedone LV4 EC (2.0–4.2 pt) Barrage HF (2–3 pt) others	2,4-D (≤1.5)	many broadleaf weeds including dandelion	annual bluegrass, bermudagrass, creeping bentgrass (spot treatment only), fine fescue, Kentucky bluegrass, perennial ryegrass, tall fescue, zoysiagrass	Amine formulations are nonvolatile and safer than ester formulations around ornamental trees and shrubs. Ester formulations provide better weed control in early spring and good control later in spring (see *Should I use an amine or ester formulation for postemergence broadleaf weed control?*, page 63). Do not allow spray drift to contact the foliage of landscape ornamentals or vegetables. For hard-to-control perennials, combine other ingredients to improve control. Limit of 2 applications per year. Do not use on newly seeded turf until well established.
2DQ (1–3 pt) NY	2,4-D + quinclorac + dicamba (see label)	broadleaf weeds	creeping bentgrass, Kentucky bluegrass, perennial ryegrass, tall fescue, zoysiagrass	Controls a broader spectrum of weeds than 2,4-D alone. Do not apply this product when temperatures are > 90°F. Limit of 2 applications per year.
4-Speed XT (3.0–4.0 pt; 1.8 pt on bentgrass)	2,4-D ester + triclopyr + dicamba + pyraflufen ethyl (see label)	broadleaf weeds	annual bluegrass, bermudagrass, creeping bentgrass, fine fescue, Kentucky bluegrass, perennial ryegrass, tall fescue, zoysiagrass	Controls a broader spectrum of broadleaf weeds than 2,4-D alone. Avoid application if air temperature exceeds 90°F. Delay reseeding 2 weeks after treatment. Limit of 2 applications per site per year. The addition of triclopyr in 4-Speed XT increases its activity on tough weed species such as ground ivy.
AAtrex 4L (12 qt) AAtrex Nine-O (1.1–2.2 lb) others W NY	atrazine (1.0–2.0)	summer annual grasses, annual bluegrass, henbit, common chickweed, some small-seeded broadleaves in warm-season turf	bermudagrass, zoysiagrass	**Do not use on cool-season turf.** Restricted use pesticide due to ground and surface water concerns. Apply only to dormant turf. Apply between Nov. 1 and Dec. 15 for winter weed control. Do not apply to areas that drain onto other desirable turfgrasses. Avoid application during spring green-up. Do not apply more than 2.0 lb ai/A for any application. Do not apply more than 3.0 lb ai/A per year.
Acclaim Extra (3.5–39.0 fl oz)	fenoxaprop (0.02–0.17)	annual grassy weeds, bermudagrass suppression	annual bluegrass, creeping bentgrass, fine fescue, Kentucky bluegrass, perennial ryegrass, tall fescue, zoysiagrass	Rate depends on growth stage of grassy weeds. Check the label. A second application may be made 2 weeks after the first. Do not use on bentgrass putting greens. Use 3.5 oz/A on creeping bentgrass. Do not exceed 28 oz/A on Kentucky bluegrass and zoysiagrass. Use the 28 oz/A rate for bermudagrass suppression. For bermudagrass suppression, tank-mixing with 32 fl oz/A of Turflon Ester Ultra will improve control.
Asulox (*sod farms only*, 5 pt) W L	asulam (2.1)	crabgrass, goosegrass, sandbur	bermudagrass	**Do not use on cool-season turf.** Make one application per season. Label lists Tifway bermudagrass but does not specify safety on other bermudagrass cultivars. Do not apply to stressed or freshly mown turf.

W This product is labeled for use primarily in warm-season turf. Read the label for a list of tolerant grasses.
NY Use or sale of this product has restrictions in New York State or Long Island, New York (Nassau and Suffolk counties). Refer to the label for more information.
L This product is registered only in specific North Central states. Check with local contacts for availability.

Postemergence Herbicides (*continued*)

Trade Name (product/A)	Common Name (lb ai or ae/A)	Weeds Controlled	Turfgrasses	Comments
Avenue South (5–8 pt) L	penoxsulam + sulfentrazone + 2,4-D + dicamba (see label)	broadleaf weeds	annual bluegrass, bermudagrass, buffalograss, fine fescue, Kentucky bluegrass, perennial ryegrass, tall fescue, zoysiagrass	Do not use on creeping bentgrass. Do not apply to golf course putting greens, tees, or collars. Do not exceed 2 applications per year. Avoid applications when warm-season grasses are transitioning into or out of dormancy. Perennial ryegrass and tall fescue are less tolerant than other turfgrasses. See label for more information.
Banvel (0.5–2.0 pt) Vanquish (0.5–2.0 pt) Diablo (0.5–2.0 pt) Clarity (*sod farms only*, 0.2–2.0 pt)	dicamba (0.25–1.0)	white clover, woodsorrel, wild onions, henbit, knotweed, lespedeza, docks, others	annual bluegrass, bermudagrass, buffalograss, creeping bentgrass, fine fescue, Kentucky bluegrass, perennial ryegrass, tall fescue, zoysiagrass	**Restricted use pesticide in some states due to crop injury concerns.** To avoid injury to newly seeded grasses, delay application until after the second mowing. Repeat applications may be needed. Do not exceed 2 pt/A per year. Dicamba is relatively mobile in the soil profile, so use care around the root zones of sensitive ornamental plants. Tank-mix with other herbicides such as 2,4-D for a broader spectrum of weed control.
Basagran T/O (2 pt)	bentazon (1.0)	yellow nutsedge, purple nutsedge, annual sedge, many broadleaf weeds including Canada thistle	annual bluegrass, bermudagrass, buffalograss, creeping bentgrass, fine fescue, Kentucky bluegrass, perennial ryegrass, tall fescue, zoysiagrass	Apply to well-established turf. Apply 2 pt/A and follow no < 21 days later with an additional application if necessary. Tank-mix a crop or seed oil concentrate at 2 pt/A for improved control. Spot treatments may result in increased injury. Do not exceed 4 pt/A per year.
Battleship III (2–4 pt) NY	MCPA + fluroxypyr + triclopyr (see label)	broadleaf weeds	bermudagrass, buffalograss, creeping bentgrass, fine fescue, Kentucky bluegrass, perennial ryegrass, tall fescue, zoysiagrass	Maximum rate on bermudagrass and zoysiagrass is 3 pt/A. On closely mown creeping bentgrass fairways, apply no more than 2 pt/A.
Blindside (3.25–10.0 oz) W L	sulfentrazone + metsulfuron (see label)	broadleaf weeds and sedges	bermudagrass, Kentucky bluegrass, tall fescue, zoysiagrass	**Use caution when using this product on cool-season grasses,** especially tall fescue, which is more sensitive to this herbicide than Kentucky bluegrass. Controls a wide range of broadleaf weeds, sedges, and wild garlic/onion. Do not add a surfactant unless previous experience indicates this will not increase turf injury.
Brushmaster (4–6 pt) Super Trimec (2–3 pt)	2,4-D + dichlorprop (2,4-DP) + dicamba (see label)	broadleaf weeds	annual bluegrass, fine fescue, Kentucky bluegrass, perennial ryegrass, tall fescue	Controls a broader spectrum of broadleaf weeds than 2,4-D alone. Avoid applications if air temperatures exceed 85°F. Limit of 2 applications per year.
Buctril (1–2 pt) Buctril 4EC (0.5–1.0 pt) others	bromoxynil (0.25–0.5)	small (seedling) broadleaf weeds	bermudagrass, creeping bentgrass (seed and sod farms only), fine fescue, Kentucky bluegrass, perennial ryegrass, tall fescue, zoysiagrass	Not for use on residential, schoolyard, or playground turf. Safe on seedling or established turf. Control is poor after weeds exceed the 4- to 6-leaf stage. Do not apply with a backpack or hand-held application equipment.
Cavalcade PQ (1.0–2.3 lb) NY	prodiamine + quinclorac (see label)	crabgrass, barnyardgrass, foxtail, white clover, dandelion, other broadleaf weeds	established bermudagrass, buffalograss, creeping bentgrass, Kentucky bluegrass, perennial ryegrass, tall fescue, zoysiagrass	Apply only to well-established turf. Do not apply to golf course putting greens or collars. Do not exceed 4.6 lb/A per year. When a sequential postemergence application is made for annual grassy weed or broadleaf control, use a methylated seed oil (1.5 pt/A) as a surfactant.

W This product is labeled for use primarily in warm-season turf. Read the label for a list of tolerant grasses.
NY Use or sale of this product has restrictions in New York State or Long Island, New York (Nassau and Suffolk counties). Refer to the label for more information.
L This product is registered only in specific North Central states. Check with local contacts for availability.

Postemergence Herbicides (*continued*)

Trade Name (product/A)	Common Name (lb ai or ae/A)	Weeds Controlled	Turfgrasses	Comments
Celero (8–14 oz) NY	imazosulfuron (0.38–0.66)	suppresses and controls kyllinga, purple nutsedge, yellow nutsedge; also controls various broadleaf weeds	bermudagrass, creeping bentgrass, fine fescue, Kentucky bluegrass, perennial ryegrass, tall fescue, zoysiagrass	Add NIS at 0.25% (v/v). More than 1 application may be needed for complete weed control. Repeat application 21 days after the initial application if needed. Do not apply to moist or wet turf. Do not apply to golf course putting greens.
Celsius WG (2.5–4.9 oz) W NY	thiencarbazone + iodosulfuron + dicamba (see label)	broadleaf weeds and grasses	bermudagrass, buffalograss, zoysiagrass	**Do not use on cool-season turf.** This product provides effective control of many broadleaf and grassy weeds in bermudagrass, buffalograss, and zoysiagrass. Do not use more than 7.4 oz/A per year.
Celsius XTRA (5–10 oz/A) W	thiencarbazone + iodosulfuron + halosulfuron (see label)	broadleaf weeds, some annual and perennial grasses, and sedges and kyllinga	bermudagrass, buffalograss, zoysiagrass	**Do not use on cool-season turf.** Do not apply where runoff onto cool-season grasses may occur. Add NIS at 0.25% to 0.5% (v/v) unless air temperature exceeds 90°F coupled with high humidity at the time of application. Do not use more than 15 oz/A per year.
Certainty (1.25–2.0 oz) W	sulfosulfuron (0.058–0.092)	annual bluegrass, yellow nutsedge, purple nutsedge, annual sedge, ryegrass, tall fescue, rough bluegrass, johnsongrass, wild garlic	bermudagrass, buffalograss, zoysiagrass	**Do not use on cool-season turf.** Add NIS at 0.25% (v/v). May be used on commercial and residential turf. Make a second application after 30 days if needed. Do not apply within 4 feet of golf course putting greens. A label change in 2011 no longer allows for use on cool-season turf. Product packaged before these changes can be used according to directions on the affixed label.
Change Up (1.25 to 3.0 pt) NY	MCPA + fluroxypyr + dicamba (see label)	broadleaf weeds	annual bluegrass, bermudagrass, creeping bentgrass, fine fescue, Kentucky bluegrass, perennial ryegrass, tall fescue, zoysiagrass	Change Up is an option for any accounts that may be sensitive to 2,4-D use on the property. Use 1.25 pt/A on creeping bentgrass. Do not use on golf course putting greens or tees. Do not exceed 2 applications per year.
Amine Formulations Chaser 2 Amine (1–2 qt) Turflon II Amine (2.5–3.5 pt) **Ester Formulations** Chaser (1–2 qt)	2,4-D + triclopyr (see label)	broadleaf weeds	annual bluegrass, creeping bentgrass, Kentucky bluegrass, perennial ryegrass, tall fescue	Controls a broader spectrum of weeds than 2,4-D alone. Not recommended for warm-season grasses. Injures creeping bentgrass. Mow newly seeded turf 2–3 times before application. Do not reseed until 3 weeks after application. Limit of 2 applications per year.
Chaser Ultra 2 (2 to 3 pt) NY L	MCPA + fluroxypyr + dichlorprop (2,4-DP)	broadleaf weeds	annual bluegrass, bermudagrass, creeping bentgrass, fine fescue, Kentucky bluegrass, perennial ryegrass, tall fescue, zoysiagrass	Controls a broader spectrum of weeds than MCPA alone. Do not use on creeping bentgrass putting greens or golf course tees. Do not exceed 2 applications per year.
Coastal (48 to 64 fl oz) W L NY	simazine + prodiamine + imazaquin (see label)	preemergence and postemergence control of many weeds	bermudagrass, zoysiagrass	**Do not use on cool-season turf.** Do not apply when turf is emerging from winter dormancy. Do not reseed or winter overseed for 4 months after treatment. Do not apply to golf course putting greens. May be applied twice per year with a maximum annual use rate of 113 fl oz/A.

W This product is labeled for use primarily in warm-season turf. Read the label for a list of tolerant grasses.
NY Use or sale of this product has restrictions in New York State or Long Island, New York (Nassau and Suffolk counties). Refer to the label for more information.
L This product is registered only in specific North Central states. Check with local contacts for availability.

Postemergence Herbicides (*continued*)

Trade Name (product/A)	Common Name (lb ai or ae/A)	Weeds Controlled	Turfgrasses	Comments
Confront (1-2 pt) 2-D (1-2 pt)	triclopyr + clopyralid (see label)	broadleaf weeds, good on legumes, plantain, clover	bermudagrass (*not sod farms*), buffalograss, creeping bentgrass, fine fescue, Kentucky bluegrass, perennial ryegrass, tall fescue, zoysiagrass	**Do not use on residential turf.** Use 1 pt/A on warm-season grasses and creeping bentgrass, and 2 pt/A on other cool-season grasses. Expect some yellowing on warm-season grasses. Use on grasses mowed at 0.5 inch or greater. Not for use on bermudagrass sod farms. Do not apply near susceptible ornamental or vegetable plants or to exposed ornamental roots. Do not collect clippings from treated areas.
Amine Formulations Horsepower (2–3 pt) Eliminate (2–3 pt) **Ester Formulations** Cool Power (2.5–3.5 pt) Three-Way Ester II (2.5–3.5 pt)	MCPA + triclopyr + dicamba (see label)	broadleaf weeds	annual bluegrass, bermudagrass, creeping bentgrass, fine fescue, Kentucky bluegrass, perennial ryegrass, tall fescue, zoysiagrass	Controls a broader spectrum of weeds than 2,4-D alone. Cool Power is an ester formulation of MCPA + triclopyr + dicamba primarily for use during the cooler months. HorsePower contains amine formulations of these herbicides. Delay reseeding 3–4 weeks after application. Avoid applications during the spring transition of warm-season turfgrasses. Use lower rates on warm-season grasses.
Chlorsulfuron 75 and Telar XP (roadside and industrial sites, see label)	chlorsulfuron (0.046–0.25)	broadleaves, wild garlic, tall fescue, perennial ryegrass	bermudagrass, buffalograss, creeping bentgrass, fine fescue, Kentucky bluegrass	An older formulation labeled for turf called Corsair is no longer sold, but existing supplies may exist. Use the high rate for tall fescue control. Very slow acting. Add NIS at 0.25% (v/v). Controls prostrate spurge. This product is also labeled for wild garlic/onion control.
Defendor SC (4 fl oz) NY	florasulam (0.013)	broadleaves such as catchweed bedstraw, common chickweed, dandelion, prickly lettuce, white clover, wild carrot, others	bermudagrass, creeping bentgrass, fine fescue, Kentucky bluegrass, perennial ryegrass, tall fescue, zoysiagrass	Defendor has soil and foliar activity at cool temperatures and provides fall and early spring postemergence broadleaf weed control. This herbicide also prevents flowering in some broadleaves (including dandelion) when applied at least 2 weeks before flowering in spring. Add NIS at 0.25% (v/v) when applied alone.
Dismiss (4–12 fl oz) Spartan 4F (*sod farms only*, 4–12 fl oz) NY	sulfentrazone (0.125–0.375)	dandelion, henbit, clovers, chickweed, spurges, speedwells, wild garlic, annual sedge, purple nutsedge, yellow nutsedge	bermudagrass, buffalograss, creeping bentgrass, fine fescue, Kentucky bluegrass, perennial ryegrass, tall fescue, zoysiagrass	Maximum use rate on bentgrass is 4 fl oz/A. Use rate on other cool-season grasses is 4–8 fl oz/A. Bermudagrass and zoysiagrass may be treated with 8–12 fl oz/A. Weed control spectrum increases when tank-mixed with other herbicides. Do not apply with surfactants. Wait 30 days after application to seed—product inhibits establishment. Do not use on newly seeded turfgrasses until they have been mowed at least twice. Avenue South, Blindside, Dismiss NXT, Echelon, Q4 Plus, Solitare, Spartan Charge, Surge, and TZONE also contain sulfentrazone.
Dismiss NXT (5.1–15.25 fl oz) Spartan Charge (sod farms only, 5.1–11.5 fl oz) NY	carfentrazone + sulfentrazone (see label)	suppresses and controls annual sedges, purple nutsedge, yellow nutsedge; also controls various broadleaf weeds	bermudagrass, buffalograss, creeping bentgrass, fine fescue, Kentucky bluegrass, perennial ryegrass, tall fescue, zoysiagrass	Adding a surfactant is not recommended, because it might result in temporary discoloration. Good coverage is needed for optimum control. A second application may be required for best control of sedges.
Drive XLR8 (0.7–2 qt) Quinclorac 1.5L (0.7–2.0 qt) Quinclorac 75DF (0.33–1.0 lb) others NY	quinclorac (0.25–0.75)	crabgrass, barnyardgrass, foxtail, white clover, dandelion	annual bluegrass, bermudagrass, buffalograss, creeping bentgrass, fine fescue, Kentucky bluegrass, perennial ryegrass, tall fescue, zoysiagrass	Not labeled for golf course putting greens or collars. Use lower rates on creeping bentgrass and fine fescues. Add COC (2 pt/A) or MSO (1.5 pt/A) to increase performance. Generally safe to use on newly seeded turf, although adding surfactant is not recommended on new seedlings. See label for details. Momentum Q, Q4 Plus, Quincept, and Solitare also contain quinclorac.

NY Use or sale of this product has restrictions in New York State or Long Island, New York (Nassau and Suffolk counties). Refer to the label for more information.

Postemergence Herbicides (*continued*)

Trade Name (product/A)	Common Name (lb ai or ae/A)	Weeds Controlled	Turfgrasses	Comments
Echelon 4SC (8–36 fl oz) Echelon G available on fertilizer carrier (see label) NY	sulfentrazone + prodiamine (see label)	provides preemergence control of many grassy weeds and provides postemergence control of chickweed, spurge, goosegrass, and sedge control	bermudagrass, buffalograss, creeping bentgrass, fine fescue, Kentucky bluegrass, perennial ryegrass, tall fescue, zoysiagrass	Apply only to well-established turf. Do not apply to golf course tees or putting greens. Do not exceed the maximum label rate for the turf species per calendar year. Allow at least 60 days between treatments. Use a split application for goosegrass control. Wait 3–6 months after application before reseeding.
Escalade 2 (2–3 pt) NY	2,4-D + fluroxypyr + dicamba (see label)	controls many broadleaf weeds	annual bluegrass, bermudagrass, creeping bentgrass, fine fescue, Kentucky bluegrass, perennial ryegrass, tall fescue, zoysiagrass	Controls a broader spectrum of weeds than 2,4-D alone. Do not use on creeping bentgrass putting greens or tees. Delay applications to newly sodded areas 3–4 weeks after planting. Apply to newly seeded areas only after well established. Limit of 2 applications per year.
Fahrenheit (3–12 oz) W	dicamba + metsulfuron (see label)	broadleaf weeds, ryegrasses	bermudagrass, buffalograss, zoysiagrass	**Do not use in cool-season turf.** Controls a wide range of broadleaf weeds, ryegrasses, and wild garlic/onion. Add NIS at 0.25% (v/v). Do not plant woody ornamentals in treated areas for 1 year after application.
Finale (2–6 qt) W NY	glufosinate (0.5–1.5)	existing vegetation	nonselective, kills existing turf; can be used on winter dormant bermudagrass, buffalograss, and zoysiagrass	**Do not use on cool-season turf.** Use 2–4. fl oz/gal of water for spot applications, and 2–6 qt/A for broadcast applications. Has limited translocation. Spray foliage on a spray-to-wet basis. Adding 8.5–17.0 lb of ammonium sulfate (spray grade) per 100 gal of water may improve control.
Foundation (3.25–4.0 pt) NY	2,4-D + triclopyr + dicamba + sulfentrazone (see label)	controls many broadleaf weeds	annual bluegrass, Kentucky bluegrass, perennial ryegrass, fine fescue, tall fescue	Controls a broader spectrum of weeds than 2,4-D alone. Not for use on putting greens or tees. Do not apply when temperatures are > 90°F. Provides yellow nutsedge suppression only. For newly seeded areas, delay application until after second mowing. Do not use adjuvants.
Fusilade II (3–16 fl oz) Ornamec 170 (see label)	fluazifop (0.05–0.25)	bermudagrass suppression, crabgrass, goosegrass, dallisgrass	use only on established fine fescue, tall fescue, or zoysiagrass	Both Fusilade II and Ornamec products can be used in landscapes and turf areas. Apply Fusilade II at 3–5 fl oz/A on zoysiagrass, 5–6 fl oz/A on tall fescue, and 8–16 fl oz/A on fine fescue. Start treatments for bermudagrass suppression in zoysiagrass around June 1 when zoysiagrass is growing well. Repeat every 30 days unless zoysiagrass has not adequately recovered from the previous application. Bermudagrass suppression in tall fescue is best attempted in the spring when bermudagrass is breaking dormancy or in the fall, prior to frost. Tall fescue may be temporarily discolored and thinning often results if applied in late spring or summer. Tank-mixing fluazifop with 32 fl oz/A of Turflon Ester Ultra will improve control. Use caution when working near sensitive grasses.
GameOn (3–4 pt) NY	2,4-D + fluroxypyr + halauxifen-methyl (see label)	controls many broadleaf weeds	bermudagrass, creeping bentgrass, fine fescue, Kentucky bluegrass, perennial ryegrass, tall fescue, zoysiagrass	Do not use on residential turf. Controls a broader spectrum of weeds than 2,4-D alone. Do not apply to golf course putting greens or bermudagrass or creeping bentgrass <0.5 inch tall. Delay seeding for 3 weeks after application. Apply to turf only after well established. Do not collect grass clippings for use in mulch or compost until 2 weeks after application. Limit of 2 broadcast applications per year, excluding spot treatments.
Image 70DG (8.6–11.4 oz) W L	Imazaquin (0.37–0.5)	perennial ryegrass, tall fescue, sedges, certain broadleaves	bermudagrass, zoysiagrass	Do not use in cool-season turf. Add NIS (0.25% = 1.0 qt/100 gal). Do not apply to golf course putting greens. Use only on well-established turf. Do not apply when turf is emerging from winter dormancy.
Katana (0.5–3.0 oz) W NY	flazasulfuron (0.008–0.047)	annual bluegrass, perennial ryegrass, tall fescue, various broadleaf and grassy weeds, sedges	bermudagrass, buffalograss, zoysiagrass	**Do not use in cool-season turf.** Add NIS at 0.25% (v/v) to the spray mix. Irrigate lightly after application. Can track from one area to another so use with caution in warm-season turf that borders cool-season turf. Katana is a sulfonylurea herbicide that is effective for cool-season grass control. It is one of the better products for tall fescue, annual bluegrass, and clump ryegrass removal in warm-season turf.

W This product is labeled for use primarily in warm-season turf. Read the label for a list of tolerant grasses.
NY Use or sale of this product has restrictions in New York State or Long Island, New York (Nassau and Suffolk counties). Refer to the label for more information.
L This product is registered only in specific North Central states. Check with local contacts for availability.

Postemergence Herbicides (*continued*)

Trade Name (product/A)	Common Name (lb ai or ae/A)	Weeds Controlled	Turfgrasses	Comments
Kerb 50WP (1–4 lb) Kerb SC T&O (1.25–5.0 pt) W	pronamide (0.5–2.0)	annual bluegrass, perennial ryegrass, and other grassy and broadleaf weeds in warm-season turf	bermudagrass, buffalograss, zoysiagrass	**Do not use in cool-season turf. Restricted use pesticide.** Preemergence and postemergence control in established turf. Works slowly (3–5 weeks). Do not apply to cool-season grasses or slopes that may drain onto cool-season grasses. A light rain or irrigation is needed to activate. Irrigate if no rain falls within 24–48 hours of application. Limit of 1 application per season.
Last Call (3.5 to 4.0 pt) NY	fenoxaprop + fluroxypyr + dicamba (see label)	broadleaf weeds, some annual and perennial grasses	fine fescue, Kentucky bluegrass, perennial ryegrass, tall fescue, zoysiagrass	Tank-mixing this herbicide with 2,4-D may reduce the effectiveness of this herbicide. Do not apply this product within 21 days following a 2,4-D application, or 5 days before a 2,4-D application. Do not use on desirable bermudagrass or creeping bentgrass. Do not use on sod farms.
LockUp (various granular products) Sapphire (0.25–1.5 pt) L	penoxsulam (see label)	broadleaf weeds, including white clover and English lawn daisy	annual bluegrass, bermudagrass, Kentucky bluegrass, perennial ryegrass, fine fescue, tall fescue, zoysiagrass; species safety varies by product (see label)	Not for use on putting greens or tees. Do not collect grass clippings for use as mulch around plants. A slight yellowing of turfgrass may occur within 2 weeks of application on tall fescue and perennial ryegrass. Tall fescue and perennial ryegrass are more prone to injury than Kentucky bluegrass.
LockUp Extra 2 (various granular products) (150–175 lb)	penoxsulam + 2,4-D + dicamba (see label)	broadleaf weeds	annual bluegrass, bermudagrass, buffalograss, Kentucky bluegrass, perennial ryegrass, fine fescue, tall fescue, zoysiagrass	Not for use on putting greens or tees. Do not collect grass clippings for use as mulch around plants. A slight yellowing of turfgrass may occur within 2 weeks of application on tall fescue and perennial ryegrass. Tall fescue and perennial ryegrass are more prone to injury than Kentucky bluegrass.
Lontrel (0.25–1.33 pt) NY	clopyralid (0.09–0.5)	broadleaf weeds including legumes and composite family weeds	bermudagrass, buffalograss, creeping bentgrass, Kentucky bluegrass, fine fescues, perennial ryegrass, tall fescue, zoysiagrass	Do not use on home lawns. Excellent legume control, including clovers. Avoid contact with any leguminous landscape plants including honey locust, mimosa, linden, and redbud. Do not collect grass clippings for mulch or compost.
Manor (0.25–1 oz) Mansion (0.25–1 oz) MSM Turf (0.25–1 oz) MSM 250D (0.56–4.47 fl oz/A)	metsulfuron (see label)	broadleaf weeds, some annual and perennial grasses	bermudagrass, buffalograss, Kentucky bluegrass, zoysiagrass	**Do not use in cool-season turf except Kentucky bluegrass.** Controls a wide range of broadleaf weeds and wild garlic/onion. May also be used to hasten the spring transition of bermudagrass overseeded with perennial ryegrass to bermudagrass. Add NIS at 0.25% (v/v). Do not exceed 0.5 oz/A on Kentucky bluegrass. Do not plant woody ornamentals in treated areas for 1 year after application.
Manuscript (9.6–19.2 fl oz) L	pinoxaden (0.03–0.06) + cloquintocet-mexyl	crabgrass, dallisgrass, ryegrass, thin paspalum	annual bluegrass, bermudagrass, fine fescue, zoysiagrass	Specialty herbicide for specific grass control within certain turfgrass species. Use as directed for spot treatments to targeted weeds. Tank-mix with Adigor surfactant or MSO at 0.5% to 1.0% (v/v) to increase performance. Cloquintocet-mexyl is an herbicide safener and is not listed on the product label as an active ingredient but is listed as an ingredient on the SDS.
MCPA-4 Amine (1–4 pt) MCPA Ester 4 (1–3 pt)	MCPA (0.46–1.85)	broadleaf weeds	annual bluegrass, bermudagrass, fine fescues, Kentucky bluegrass, perennial ryegrass, tall fescue, zoysiagrass	Apply in spring and fall for best results. Do not use on newly seeded turf until well established. Do not apply if rainfall is expected within 48 hours.
Mecomec 2.5 (4–6 pt) Mecomec 4 (2.5–4 pt)	mecoprop-p (MCPP) (1.0–2.0)	broadleaf weeds	annual bluegrass, bermudagrass, creeping bentgrass, Kentucky bluegrass, perennial ryegrass, tall fescue, zoysiagrass	Use only on actively growing turf that is not under stress. Do not apply to bentgrass in the heat of summer. Use low rate on putting greens.
Millennium Ultra 2 (2–3 pt) NY	2,4-D + clopyralid + dicamba (see label)	broadleaf weeds	annual bluegrass, bermudagrass, creeping bentgrass, fine fescue, Kentucky bluegrass, perennial ryegrass, tall fescue, zoysiagrass	Not labeled for use on residential turf. Controls a broader spectrum of weeds than 2,4-D alone. Delay application to newly seeded grasses until after 3–4 mowings. Delay reseeding 3–4 weeks after application. Do not spray when air temperatures are > 90°F. Not recommended for use on creeping bentgrass putting greens and tees. Limit of 2 applications per year. Do not collect clippings from treated areas.

W This product is labeled for use primarily in warm-season turf. Read the label for a list of tolerant grasses.
NY Use or sale of this product has restrictions in New York State or Long Island, New York (Nassau and Suffolk counties). Refer to the label for more information.
L This product is registered only in specific North Central states. Check with local contacts for availability.

Postemergence Herbicides (*continued*)

Trade Name (product/A)	Common Name (lb ai or ae/A)	Weeds Controlled	Turfgrasses	Comments
Momentum 4-Score (3.5–4.5 pt) NY	2,4-D + fluroxypyr + triclopyr + sulfentrazone (see label)	broadleaf weeds	annual bluegrass, creeping bentgrass, Kentucky bluegrass, perennial ryegrass, fine fescue, tall fescue	Controls a broader spectrum of weeds than 2,4-D alone. Not for use on putting greens or tees. Do not apply when temperatures are > 90°F. Provides yellow nutsedge suppression only. For newly seeded areas, delay application until after second mowing. Do not use adjuvants.
Momentum FX2 (3–4 pt) NY	2,4-D + triclopyr + fluroxypyr (see label)	broadleaf weeds	bermudagrass, creeping bentgrass, fine fescue, Kentucky bluegrass, perennial ryegrass, tall fescue, zoysiagrass	Controls a broader spectrum of weeds than 2,4-D alone. Use the 3 pt/A rate on warm-season grasses. Mow newly seeded areas twice before application. Do not reseed within 3-4 weeks after application. Avoid applications during the spring transition of warm-season turfgrasses. Limit of 2 applications per year.
Momentum Q (7–8 pt) Quincept (7–8 pt) NY	2,4-D + quinclorac + dicamba (see label)	crabgrass, broadleaf weeds	annual bluegrass, bermudagrass, creeping bentgrass, fine fescue, Kentucky bluegrass, perennial ryegrass, tall fescue, zoysiagrass	Controls a broader spectrum of weeds than 2,4-D alone and controls some summer annual grasses. Yellowing may occur on creeping bentgrass. Use 3–5 pt/A on creeping bentgrass. Delay application to newly seeded turf until after the third mowing. Do not apply this product when temperatures are > 90°F. Limit of 2 applications per year.
Monument 75 WG (0.35–0.53 oz) W L NY	trifloxysulfuron (0.016–0.025)	suppresses and controls annual sedges, purple nutsedge, and yellow nutsedge; controls annual bluegrass, ryegrass, tall fescue, and certain broadleaves	bermudagrass, buffalograss, zoysiagrass	**Do not use on cool-season turf.** Do not apply where runoff onto cool-season grasses may occur. Add NIS at 0.25% (v/v). Repeat application when regrowth appears. To reduce potential movement, water-in lightly 2–3 hours after application to remove product from turf foliage. Allow turf to dry before allowing traffic onto treated areas.
MSMA 6 Plus (2.7 pt) MSMA 6.6 (2.5–5.0 pt) Target 6 Plus (2.7 pt) Target 6.6 (2.7 pt) L	MSMA (2.0–4.4)	crabgrass, goosegrass, yellow nutsedge, dallisgrass, sandbur	annual bluegrass, bermudagrass, Kentucky bluegrass, zoysiagrass	Temporary discoloration of turf will occur. Sedges, field paspalum, and sandbur control may require multiple applications for control. **MSMA use is being phased out by the EPA**—see *Can I still use MSMA?,* page 66 for details. Sale prohibited to residential turf including athletic fields, parks, and commercial turf after Dec. 31, 2010. Only spot treatments (100 ft^2 maximum per spot) are allowed on existing golf courses. Sod farms can make 2 broadcast applications per year. Sod farms and golf courses only.
NativeKlean (19–34 fl oz) NY	2,4-D + aminopyralid (see label)	broadleaf weeds including invasive or noxious weeds (see label)	native or natural areas that are not regularly mowed or maintained	Controls a broader spectrum of difficult-to-control broadleaf weeds. Do not apply this product on managed turf or sod farms. Apply no more than 2.1 pints (34 fluid oz).
Negate 37WG (1.5 oz) W L	metsulfuron + rimsulfuron (0.019 + 0.015)	broadleaf weeds, some annual and perennial grasses	bermudagrass, zoysiagrass	**Do not use on cool-season turf.** May be used to control emerged weeds in bermudagrass and zoysiagrass. Add NIS at 0.25% (v/v). Apply to dormant or nondormant turf. Some chlorosis or stunting of turf may occur after application. Use a 15-foot buffer area around cool-season grasses.
Octane 2% SC (0.7–4.0 fl oz)	pyraflufen ethyl (0.001–0.006)	chickweed, white clover, dandelion	bermudagrass, creeping bentgrass, Kentucky bluegrass, perennial ryegrass, tall fescue, zoysiagrass	Can treat newly seeded areas not under moisture, temperature, or mechanical stress. Treated areas can be overseeded 1 day after application. Causes rapid necrosis of sensitive weeds. Tank-mix with other herbicides for best weed control.
On Deck (1.0 to 4.0 pt)	2,4-D + dicamba (see label)	broadleaf weeds	bermudagrass, buffalograss, creeping bentgrass, Kentucky bluegrass, perennial ryegrass, tall fescue, zoysiagrass	Controls a broader spectrum of weeds than 2,4-D alone. Do not apply more than 4 pt/A per application to cool-season grasses, or more than 1.5 pt/A per application to warm-season grasses. Limit of 2 applications per year.
Patron 170 (3.5 qt) NY	2,4-D + dichlorprop (2,4- DP) (see label)	broadleaf weeds	annual bluegrass, bermudagrass, fine fescue, Kentucky bluegrass, perennial ryegrass, tall fescue, zoysiagrass	Controls a broader spectrum of weeds than 2,4-D alone. Fall-seeded lawns may be treated the following spring. Spring-seeded lawns may be treated after 2 mowings. This product may injure creeping bentgrass. Do not use on golf course putting greens.

W This product is labeled for use primarily in warm-season turf. Read the label for a list of tolerant grasses.
NY Use or sale of this product has restrictions in New York State or Long Island, New York (Nassau and Suffolk counties). Refer to the label for more information.
L This product is only registered in specific North Central states. Check with local contacts for availability.

Trade Name (product/A)	Common Name (lb ai or ae/A)	Weeds Controlled	Turfgrasses	Comments
Plateau (2 to 6 fl oz) W NY	imazapic (0.03–0.09)	summer annual and perennial grasses, some broadleaf weeds, sedges	bermudagrass, buffalograss	**Do not use on cool-season turf.** Tolerance to this herbicide varies by cultivar. Rates of 2–6 fl oz/A will control some weeds. Some bermudagrass seedheads also will be suppressed.
PowerZone (2–6 pt)	MCPA + MCPP + dicamba + carfentrazone (see label)	many broadleaf weeds	annual bluegrass, bermudagrass, fine fescue, Kentucky bluegrass, perennial ryegrass, tall fescue, zoysiagrass	Controls a wide range of broadleaf weeds. May be applied after the second mowing in newly established cool-season turfgrasses. Do not reseed until 2 weeks after application. Do not apply when temperatures are > 90°F.
Princep Liquid (1–2 qt) Simazine 90DF (1.1–2.2 lb) others W	simazine (1.0–2.0)	summer annual grasses, annual bluegrass, henbit, common chickweed, some small-seeded broadleaves in warm-season turf	established bermudagrass, buffalograss, zoysiagrass	**Do not use on cool-season turf.** Apply only to dormant bermudagrass and zoysiagrass. Apply between Nov. 1 and Dec. 15 for winter weed control. Do not apply to areas that drain onto golf course tees, putting greens, or other desirable turfgrasses. Avoid application during spring green-up. Do not exceed 3 qt/A per calendar year. Make no more than 2 applications per year.
PoaCure (0.2–0.4 gal) NY	methiozolin (0.46–0.92)	preemergence and postemergence control of annual bluegrass; postemergence rough bluegrass control	bermudagrass, creeping bentgrass, fine fescue, Kentucky bluegrass, perennial ryegrass, tall fescue, zoysiagrass	**For use on golf courses only.** Water-in sprays with 0.1 to 0.2-inch of irrigation after application. The herbicide is slow acting, taking 2 to 4 weeks to control weeds. Do not treat putting greens under stress or within 1 week of aerification. Check the label for application timing, rate, and frequency recommendations.
Prograss (0.5–1.33 gal) Prograss SC (0.75–2.0 qt)	ethofumesate (0.75–2.0)	annual bluegrass, crabgrass, and white clover	creeping bentgrass, bermudagrass (dormant), perennial ryegrass, Kentucky bluegrass, tall fescue	See label for recommendations for reducing annual bluegrass in cool-season turf. Do not use more than 0.5 gal/A on Kentucky bluegrass or creeping bentgrass. Delay seeding Kentucky bluegrass until 6 weeks after the last Prograss application. Refer to the label for further instructions on each grass species.
Pylex (1.0–1.5 fl oz)	topramezone (0.02–0.03)	bermudagrass, goosegrass, nimblewill, windmillgrass, summer annual grasses, broadleaves	fine fescue, Kentucky bluegrass, perennial ryegrass, tall fescue	Has a bleaching effect on susceptible species. Do not apply more than 4 fl oz (0.089 lb ai)/A per year. Add COC or MSO for best results at 0.5%–1.0% (v/v). Make a repeat application 3 weeks after the first. Apply prior to seeding, at seeding, or 28 days or more after seeding.
Q4 Plus (5–8 pt) NY	2,4-D + quinclorac + dicamba + sulfentrazone (see label)	controls broadleaf weeds and some annual grasses such as crabgrass; also suppresses and controls annual sedges, purple nutsedge, and yellow nutsedge	annual bluegrass, bermudagrass, fine fescue, Kentucky bluegrass, perennial ryegrass, tall fescue, zoysiagrass	Do not use on creeping bentgrass. Do not apply to golf course putting greens, tees, or collars. Make no more than 2 applications per year. Avoid applications when warm-season grasses are transitioning into or out of dormancy. Q4 is an older product with the same ingredients but 33% less quinclorac.
Quicksilver T&O (1.0–6.7 fl oz) Aim EC (*sod farms only,* 0.5–2.0 fl oz)	carfentrazone (see label)	some broadleaf weeds (1.0–2.1 fl oz) and silvery thread moss (2.0–6.7 fl oz)	annual bluegrass, bermudagrass, buffalograss, creeping bentgrass, Kentucky bluegrass, perennial ryegrass, tall fescue, zoysiagrass	Tall fescue may experience some short-term yellowing after application. Can be applied 7 days after emergence (after seeding or sprigging) of bermudagrass, creeping bentgrass, tall fescue, perennial ryegrass, and Kentucky bluegrass. Treated areas can be seeded 1 day after application. Use NIS at 0.25% (v/v). Causes rapid necrosis of sensitive weeds. Tank-mix with other herbicides for best weed control.
Recognition (0.37–1.95 oz/A) W L NY	trifloxysulfuron + metcamifen	suppresses and controls sedges, annual bluegrass, ryegrasses, tall fescue, and certain broadleaf weeds	bermudagrass, buffalograss, zoysiagrass	**Do not use on cool-season turf.** Add NIS at 0.25% (v/v). Do not apply where runoff onto cool-season grasses may occur. Repeat application when regrowth appears. To reduce potential movement, water-in lightly 2–3 hours after application to remove product from turf foliage. Allow turf to dry before allowing traffic onto treated areas. Tank-mix with other herbicides to improve the spectrum of control. A 0.53 oz/A rate of Monument is equivalent trifloxysulfuron to a 1.95 oz/A rate of Recognition. Metcamifen is an herbicide safener and is not listed on the product label as an active ingredient but is listed as an ingredient on the SDS.

W This product is labeled for use primarily in warm-season turf. Read the label for a list of tolerant grasses.
NY Use or sale of this product has restrictions in New York State or Long Island, New York (Nassau and Suffolk counties). Refer to the label for more information.
L This product is only registered in specific North Central states. Check with local contacts for availability

Trade Name (product/A)	Common Name (lb ai or ae/A)	Weeds Controlled	Turfgrasses	Comments
RedZone 2 (3.0–4.0 pt; 1.8 pt on bentgrass)	2,4-D ester + MCPP + dicamba + pyraflufen ethyl (see label)	broadleaf weeds	annual bluegrass, bermudagrass, creeping bentgrass, fine fescue, Kentucky bluegrass, perennial ryegrass, tall fescue, zoysiagrass	Controls a broader spectrum of broadleaf weeds than 2,4-D alone. Avoid applications if air temperatures exceed 90°F. Delay reseeding 2 weeks after treatment. Limit of 2 applications per site per year.
Revolver (4.4–35.2 fl oz) W NY	foramsulfuron (0.006–0.052)	annual bluegrass, goosegrass, ryegrass, tall fescue, partial dallisgrass control	bermudagrass, buffalograss, zoysiagrass	**Do not use on cool-season turf.** Apply for (1) control of undesirable cool-season grasses (including tall fescue), (2) postemergence control of goosegrass in late spring and early summer, and (3) spring transition of bermudagrass overseeded the previous fall with cool-season turfgrasses. Rainfast within 2 hours of application.
Reward (1–2 pt) Diquat SPC 2L (1–2 pt) others W	diquat (0.25–0.5)	winter annual broadleaf weeds such as annual bluegrass and henbit	dormant bermudagrass, buffalograss, zoysiagrass	**Do not use on cool-season turf.** For spot spray, add 0.75 fl oz/gal. Add NIS at 0.25% (v/v) to the spray mix. Use at least 15 gal/A spray volume. Lower water volumes (≤20 gal/A) will reduce risk of injury to dormant bermudagrass and zoysiagrass for late winter applications.
Rimsulfuron 25 DF (0.5–2.0 oz) W L	rimsulfuron (0.007–0.03)	henbit, purple deadnettle, woodsorrel, plantains, cool-season grasses	bermudagrass, zoysiagrass	**Do not use on cool-season turf.** May be used to control emerged weeds in bermudagrass and zoysiagrass. Add NIS at 0.25% (v/v). To hasten spring transition on overseeded bermudagrass, apply at the 60%–75% green-up growth stage of bermudagrass or approximately 2–3 weeks before transition is desired. Do not apply to residential lawns.
Roundup PROMAX (1–3.3.0 qt) Roundup Original MAX (0.4–3.3 qt) others W	glyphosate (see label)	annual bluegrass, winter broadleaf weeds	dormant bermudagrass buffalograss, zoysiagrass	**Do not use on cool-season turf.** Apply only to completely dormant bermudagrass with no green stolons or leaves visible. Late February or early March in Indiana, Illinois, Kansas, Kentucky, Missouri, and Ohio is usually the best time to apply. Lower water volumes (≤20 gal/A) will reduce risk of injury to dormant bermudagrass and zoysiagrass for late winter applications.
SedgeHammer or ProSedge (0.66–1.33 oz) Sandea or Profine75 (*sod farms only*, 0.66–1.33 oz) SedgeHammer+ (21 oz)	halosulfuron (see label)	yellow nutsedge, purple nutsedge	annual bluegrass, bermudagrass, buffalograss, creeping bentgrass, fine fescue, Kentucky bluegrass, perennial ryegrass, tall fescue, zoysiagrass	SedgeHammer+ already contains surfactant, so don't add it. For other products, use NIS at 0.25%–0.5% (v/v). 2 applications may be made with the second 6 weeks or later following the first application. Do not exceed 0.125 lb/ai/A in a 12-month period. Not labeled for putting greens. For spot treatment, mix 0.9 gram of 75DF-formulated halosulfuron in 1 gal of water with 0.33 oz (2 tsp) surfactant. For spot treatment with SedgeHammer+ use 0.5 oz/1,000 ft^2.
Segment II (1.0 pt) Segment (1.5 pt)	sethoxydim (0.1875)	crabgrass, goosegrass, sandbur, other grassy weeds	fine fescue	Fine fescues only. Do not make more than 2 applications per season. Add a COC or MSO when using Segment II. Do not add an adjuvant when using Segment because it already contains crop oil. Apply no sooner than 3 weeks after spring green-up. Do not apply to newly established fine fescue. Apply before crabgrass becomes well tillered. Do not mow 7 days before or after application. Product is most effective when applied in early spring and late fall.
Sencor 75% (0.33–0.66 lb) W L	metribuzin (0.25 to 0.5)	annual bluegrass, goosegrass, broadleaf annual weeds	bermudagrass	**Do not use on cool-season turf.** Use low rate on actively growing bermudagrass. Use high rate on dormant bermudagrass for winter annual weed control. Do not apply in the root zone of shallow-rooted ornamentals. Do not apply to golf course tees, putting greens, or other turf mowed at < ½ inch.

W This product is labeled for use primarily in warm-season turf. Read the label for a list of tolerant grasses.
NY Use or sale of this product has restrictions in New York State or Long Island, New York (Nassau and Suffolk counties). Refer to the label for more information.
L This product is only registered in specific North Central states. Check with local contacts for availability.

Trade Name (product/A)	Common Name (lb ai or ae/A)	Weeds Controlled	Turfgrasses	Comments
Solitaire (1–2 lb) Solitaire WSL (5.4–10.7 qt) NY	quinclorac + sulfentrazone (see label)	suppresses and controls annual sedges, purple nutsedge, and yellow nutsedge; also controls various broadleaf weeds and summer annual grasses such as crabgrass	bermudagrass, buffalograss, fine fescue, Kentucky bluegrass, perennial ryegrass, tall fescue, zoysiagrass	Do not apply to golf course putting greens, collars, or tees. Do not apply directly to landscape ornamentals or to ornamental beds. Do not apply with surfactants unless compatibility tests have previously demonstrated compatibility and safety on grass type. Seeding should be delayed 4 weeks after application. Do not use on newly seeded turfgrasses until they have been mowed at least twice. Solitaire WSL is labeled for use in fine fescues.
Specticle FLO (3–10 fl oz) Specticle G (100–200 lb) W NY	indaziflam (see label)	postemergence crabgrass control (≤1 tiller), annual bluegrass, winter annual broadleaves	established bermudagrass, buffalograss, zoysiagrass	**Do not use on cool-season turf unless thinning or removal is desired.** Do not apply to golf course putting greens, tees, or collars. Irrigate immediately after application. Use a 15-foot buffer area around cool-season grasses.
SpeedZone (2–5 pt) SpeedZone Southern (2–6 pt) L	2,4-D + MCPP + dicamba + carfentrazone (see label)	broadleaf weeds, goosegrass	annual bluegrass, bermudagrass, buffalograss, creeping bentgrass, fine fescues, Kentucky bluegrass, perennial ryegrass, tall fescue, zoysiagrass	Controls a broader spectrum of weeds than 2,4-D alone. May be applied after the second mowing in newly established cool-season turfgrasses. Do not reseed until 1 week after application. Do not apply when temperatures are > 90°F. Limit of 2 applications per year. Speedzone Southern is formulated to deliver less 2,4-D + MCPP + dicamba to increase safety on turfgrasses adapted to areas with warmer climates (that is, bahiagrass, centipedegrass, seashore paspalum, kikuyugrass).
Sublime (32 fl oz/A)	triclopyr + dicamba + mesotrione	broadleaf weeds, yellow nutsedge, some annual and perennial grasses	bermudagrass, fine fescue, Kentucky bluegrass, perennial ryegrass, tall fescue, zoysiagrass	Controls a wide range of weeds. May cause a bleaching (whitening) effect on susceptible species, but this effect is muted by the other herbicides in the mixture. **Only use on bermudagrass during winter dormancy**. Add NIS at 0.25% (v/v). The herbicide will cause short-term zoysiagrass discoloration after treatment. Do not apply more than 4 applications a year. This herbicide can be applied prior to or at seeding. Do not apply to newly germinated turfgrass seedings until they become well established.
Sure Power (2.0–3.5 pt) NY	2,4-D + triclopyr + fluroxypyr + flumioxazin (see label)	broadleaf weeds	Kentucky bluegrass, perennial ryegrass, fine fescue, tall fescue	Do not use on creeping bentgrass or turf with significant annual or roughstalk bluegrass. Controls a wide range of broadleaf weeds. Wait at least 4 weeks after applying before reseeding or sodding. Avoid broadcast applications for 14 days after freezing conditions or frost. Adding an adjuvant is not recommended and will increase turf injury risk. Applying when humidity is high will result in turfgrass injury, especially on tall fescue. This herbicide is quick acting with results visible in 2–3 days.
Surge (2.75–4.0 pt) SureZone (2.75–4.5 pt) NY	2,4-D + MCPP + dicamba + sulfentrazone (see label)	yellow nutsedge suppression; also controls various broadleaf weeds	annual bluegrass, bermudagrass, buffalograss, creeping bentgrass, fine fescue, Kentucky bluegrass, perennial ryegrass, tall fescue, zoysiagrass	Controls a broader spectrum of weeds than 2,4-D alone. Do not apply when temperatures are > 90°F or during spring transition. Provides yellow nutsedge suppression only. For newly seeded areas, delay application until after second mowing.

W This product is labeled for use primarily in warm-season turf. Read the label for a list of tolerant grasses.
NY Use or sale of this product has restrictions in New York State or Long Island, New York (Nassau and Suffolk counties). Refer to the label for more information.
L This product is only registered in specific North Central states. Check with local contacts for availability.

Postemergence Herbicides (*continued*)

Trade Name (product/A)	Common Name (lb ai or ae/A)	Weeds Controlled	Turfgrasses	Comments
Tailspin (3–6 pt) NY	triclopyr + fluroxypyr (see label)	broadleaf weeds	creeping bentgrass, fine fescue, Kentucky bluegrass, perennial ryegrass, tall fescue, zoysiagrass	Do not apply to bentgrass unless injury can be tolerated. Do not use more than 12 pt/A per year. Do not apply during adverse conditions such as drought or very hot or cold weather.
Tenacity (5–8 fl oz) TRIONE (5–8 fl oz) Meso 4SC Select (5–8 fl oz)	mesotrione (0.156–0.25)	crabgrass, creeping bentgrass, nimblewill, windmillgrass, yellow nutsedge, broadleaves	dormant bermudagrass, buffalograss, fine fescue, Kentucky bluegrass, perennial ryegrass, tall fescue	Has a bleaching effect on susceptible species. Do not apply more than 16 fl oz (0.5 lb ai)/A per year. Add NIS at 0.25% (v/v). Make a repeat application 2 weeks after the first. For best crabgrass control, apply to < 4-tiller plants. Apply prior to seeding, at seeding, or after the newly germinated turf has been mown 2 times or 4 weeks after emergence (whichever is longer). Controls nimblewill and suppresses creeping bentgrass with 3 applications at 5.3 oz/A or 4 applications at 4.0 oz/A.
Triad Select (1.8–4.0 pt/A)	2,4-D + MCPA + dicamba (see label)	controls many broadleaf weeds	annual bluegrass, bermudagrass, creeping bentgrass, fine fescues, Kentucky bluegrass, perennial ryegrass, tall fescue, zoysiagrass	Controls a broader spectrum of weeds than 2,4-D alone. Avoid applications when temperatures are > 90°F. Delay reseeding 4 weeks after treatment. Do not apply to newly seeded grasses. Limit of 2 applications per year.
Triad SFZ Select (2.75–4.0 pt/A) NY	2,4-D + MCPA + dicamba + sulfentrazone (see label)	controls many broadleaf weeds	annual bluegrass, bermudagrass, buffalograss, creeping bentgrass, fine fescues, Kentucky bluegrass, perennial ryegrass, tall fescue, zoysiagrass	Controls a broader spectrum of weeds than 2,4-D alone. Not for use on putting greens or tees. Avoid applications when temperatures are > 90°F. Delay reseeding 3 weeks after treatment. Do not apply to newly seeded grasses until mown twice. Limit of 2 applications per year. Do not use adjuvants.
Triad T Select (2–4 pt/A)	2,4-D + MCPA + triclopyr + dicamba	broadleaf weeds	annual bluegrass, bermudagrass, fine fescue, Kentucky bluegrass, perennial ryegrass, tall fescue, zoysiagrass	Controls a broader spectrum of weeds than 2,4-D alone with triclopyr, improving control of hard-to-control weeds. Triad T Select is an amine formulation. Do not apply to newly seeded grasses until well established. Avoid applications during the spring transition of warm-season turfgrasses. Use lower rates on warm-season grasses.
Triamine (1.8–4.0 pt) Spoiler (1.7–4 .0pt)	2,4-D + MCPP + dichlorprop (2,4-DP) (see label)	broadleaf weeds	annual bluegrass, bermudagrass, creeping bentgrass, fine fescues, Kentucky bluegrass, perennial ryegrass, tall fescue, zoysiagrass	Controls a broader spectrum of weeds than 2,4-D alone. Use lower rates on creeping bentgrass and warm-season turf. Do not use on creeping bentgrass putting greens or tees. Limit of 2 applications per year.
Tribute TOTAL (1.0–3.2 oz) W NY	thiencarbazone + foramsulfuron + halosulfuron (see label)	broadleaf weeds, annual and perennial grassy weeds, and sedges and kyllinga	bermudagrass, zoysiagrass	**Do not use on cool-season turf.** Do not apply where runoff onto cool-season grasses may occur. Add NIS at 0.25% to 0.5% (v/v). Repeat application when regrowth appears. See label for instructions on using Tribute TOTAL next to sensitive areas such as golf course putting greens. Allow turf to dry before allowing traffic onto treated areas. There is a yearly maximum of 6.4 oz/A.

W This product is labeled for use primarily in warm-season turf. Read the label for a list of tolerant grasses.

NY Use or sale of this product has restrictions in New York State or Long Island, New York (Nassau and Suffolk counties). Refer to the label for more information.

Postemergence Herbicides (*continued*)

Trade Name (product/A)	Common Name (lb ai or ae/A)	Weeds Controlled	Turfgrasses	Comments
Trimec Classic 3-D Eliminate-D EndRun Mec Amine-D Strike 3 Threesome Three-Way Trimec 899 Trimec 992 Trimec Bentgrass Formula Triplet Low Odor Triplet SF Vessel others (1.0–4.0 pt)	2,4-D + MCPP + dicamba (see label)	broadleaf weeds	annual bluegrass, bermudagrass, buffalograss, creeping bentgrass, fine fescue, Kentucky bluegrass, perennial ryegrass, tall fescue, zoysiagrass	Controls a broader spectrum of weeds than 2,4-D alone. Several commercial formulations with varying proportions of these ingredients are available. Refer to individual labels for recommended amounts of formulated product. Avoid applications when temperatures are > 85°F (some labels state 90°F). Delay reseeding 4 weeks after treatment. Limit of 2 applications per year.
Trimec Encore (1.8–4.0 pt) Tri-Power (2.0–4.1 pt)	MCPA + MCPP + dicamba (see label)	broadleaf weeds	annual bluegrass, bermudagrass, creeping bentgrass, fine fescue, Kentucky bluegrass, perennial ryegrass, tall fescue, zoysiagrass	Delay application to newly seeded grasses until after 3–4 mowings. Delay reseeding 3–4 weeks after application. Avoid applications during the spring transition of warm-season turfgrasses unless a slight delay in green-up can be tolerated.
Turflon Ester Ultra (0.5–1.0 qt) Triclopyr 4 (0.5–1.0 qt) Trycera (0.7–1.4 qt)	triclopyr (0.5–1.0)	broadleaf weeds, bermudagrass suppression	annual bluegrass, fine fescue, Kentucky bluegrass, perennial ryegrass, tall fescue, zoysiagrass	Repeat applications of 1 qt/A spaced 4 weeks apart are recommended for bermudagrass, corn speedwell, violets, ground ivy, and yellow woodsorrel. A previous formulation (Turflon Ester) contained a petroleum solvent adjuvant, whereas the new formulation (Turflon Ester Ultra) uses MSO.
TZONE SE (2–4 pt) TZONE (2–4 pt) Triad TZ Select (2–4 pt) NY	2,4-D + triclopyr + dicamba + sulfentrazone (see label)	broadleaf weeds, yellow nutsedge suppression	annual bluegrass, bermudagrass, fine fescue, Kentucky bluegrass, perennial ryegrass, tall fescue, zoysiagrass	Controls a broader spectrum of weeds than 2,4-D alone. Rate varies by turf species. Use 2–2.25 pt/A on bermudagrass and zoysiagrass; use 3.25–4 pt/A on cool-season turf. Limit of 2 applications per year. Delay application to newly seeded turf until after 2–3 mowings. Delay application 3–4 weeks after sodding, sprigging, or plugging turf. Treated areas may be reseeded 3 weeks after application. Do not use when temperatures are > 85°F. Do not spray when warm-season turfgrass is emerging from winter dormancy.
Vista XRT (6–23 fl oz) NY	fluroxypyr (0.13–0.50)	broadleaf weeds	bermudagrass, creeping bentgrass, fine fescue, Kentucky bluegrass, perennial ryegrass, tall fescue, zoysiagrass	Do not use on golf course putting greens or tees. Avoid application when warm-season turfgrasses are emerging from winter dormancy. Use the low application rate on creeping bentgrass and bermudagrass. Formerly, Spotlight herbicide was sold with this same ingredient. Vista XRT is a more concentrated formulation: 2.8 lb fluroxypyr ae/gal compared to 1.5 lb ae/gal in Spotlight. Tank-mixtures with other active ingredients will control a broader spectrum of weeds.
Xonerate 2SC (3–6 fl oz) NY	amicarbazone (0.04–0.09)	annual bluegrass some broadleaf weeds	bermudagrass, buffalograss, creeping bentgrass, fine fescue, Kentucky bluegrass, perennial ryegrass, tall fescue, zoysiagrass	Do not use on golf course putting greens. Apply in spring for annual bluegrass control. Do not apply in summer or fall. Apply when daily high temperatures are 55°F–80. Irrigate soon after application with 0.1 to 0.2 inches of water. Turf quality of Xonerate 2SC treated areas with >10% annual bluegrass will be temporarily reduced. Do not apply to areas where mefluidide has been applied in the previous 3 months.
Yukon (*sod farms only*, 4–8 oz)	dicamba + halosulfuron (see label)	yellow nutsedge, purple nutsedge, various broadleaf weeds	annual bluegrass, bermudagrass, creeping bentgrass, fine fescue, Kentucky bluegrass, perennial ryegrass, tall fescue, zoysiagrass	Use NIS at 0.25%–0.5% (v/v). A second application may be made 6 weeks or more after the first. Do not exceed 8 oz product/A in a single season.

NY Use or sale of this product has restrictions in New York State or Long Island, New York (Nassau and Suffolk counties). Refer to the label for more information.

Commonly Used Broadleaf Herbicide Combinations for Turfgrass

This table shows only broadleaf herbicides with more than one ingredient or herbicides that are a key ingredient in mixtures. This table compares ingredients and allows users to quickly search for products that contain a particular ingredient.

Herbicides	2,4-D	mecoprop (MCPP)	dicamba	aminopyralid	carfentrazone	clopyralid	dichlorprop (2,4-DP)	fenoxaprop	flumioxazin	fluroxypyr	halauxifen-methyl	iodosulfuron	MCPA	mesotrione	metsulfuron	penoxsulam	pyraflufen ethyl	quinclorac[1]	sulfentrazone	thiencarbazone	triclopyr
2,4-D Amine 4, Barrage HF,[2] Clean Amine, Hardball, Saber, Shredder 2,4-D LV4, Shredder 2,4-D LV6, Shredder Amine 4, Shredder E-99, WEEDestroy AM40, Weedone LV4 EC, Weedar 64	■																				
3-D, Eliminate LO, Eliminate-D, EndRun, MEC Amine-D, Strike 3, Threesome, Three-Way, Trimec 1000, Trimec 992, Trimec 899, Trimec Bentgrass Formula, Trimec Classic, Trimec LAF-637, Trimec Southern, Triplet Low Odor, Triplet SF, TruPower2, TruPower3, Vessel, others	■	■	■																		
4-Runner			■										■						■		■
4-Speed XT	■		■														■				■
Aquasweep, Chaser, Chaser 2 Amine, Crossroad, Everett, Turflon II Amine	■																				■
Avenue South	■		■													■			■		
Banvel, Clarity, Clash, Cruise Control, Detonate, Diablo, Rifle, Sterling Blue, Strut, Topeka, Vanquish, Vision			■																		
Battleship III										■			■								■
Brushmaster, Super Trimec	■		■				■														
Celsius WG			■									■								■	
Change Up			■							■			■								
Chaser Ultra 2							■			■			■								
Confront, 2-D						■															■
Cool Power, Three-Way Ester II, Horsepower, Eliminate			■										■								■
Dismiss, Spartan 4F[2]																			■		
Dismiss NXT Spartan Charge[2]					■														■		
Drive XLR8, Quinclorac 1.5L, QuinPro, Eject 75DF, others																		■			
Escalade 2	■		■							■											
Fahrenheit																					
Foundation	■		■																■		■
GameOn	■									■	■										
Last Call			■					■		■											
LockUp, Sapphire																■					
LockUp Extra 2	■		■													■					
Lontrel						■															
MCPA-4 Amine, MCPA Ester													■								
MCPP-p 4 Amine, Mecomec 2.5, Mecomec 4		■																			
Millennium Ultra 2	■		■			■															
Momentum FX2	■									■											■

Commonly Used Broadleaf Herbicide Combinations for Turfgrass (*continued*)

Herbicides	2,4-D	mecoprop (MCPP)	dicamba	aminopyralid	carfentrazone	clopyralid	dichlorprop (2,4-DP)	fenoxaprop	flumioxazin	fluroxypyr	halauxifen-methyl	iodosulfuron	MCPA	mesotrione	metsulfuron	penoxsulam	pyraflufen ethyl	quinclorac[1]	sulfentrazone	thiencarbazone	triclopyr
Momentum 4-Score	■									■									■		■
Momentum Q, Quincept, 2DQ, Triad QC Select	■		■															■			
NativeKlean	■			■																	
Octane 2% SC																	■				
Patron 170	■						■														
Power Zone		■	■		■								■								
Q4 Plus	■		■															■	■		
QuickSilver T&O, Aim EC					■																
RedZone 2	■	■	■														■				
Solitaire, Solitaire WSL																		■	■		
SpeedZone, SpeedZone Southern	■	■	■		■																
Spoiler, Triamine, Triple Threat	■	■					■														
Sublime			■											■							■
Sure Power, Tetra	■								■	■											■
Surge, SureZone	■	■	■																■		
Tailspin										■											■
Triad Select	■		■										■								
Triad T Select	■		■										■								■
Triad SFZ Select	■		■										■						■		
Tri-Power, Trimec Encore		■	■										■								
Turflon Ester Ultra, Triclopyr 4, Trycera																					■
TZONE, TZONE SE, Triad TZ Select	■		■																■		■
Vista XRT, Flagstaff										■											

1 Quinclorac (Drive) is a grass herbicide active on crabgrass and foxtails with additional activity on certain broadleaf weeds like dandelion, speedwell, clover, bindweed, and black medic.
2 For use on sod farms only.

Active Ingredients in Commonly Used Postemergence Herbicide Combinations

This table only shows broadleaf herbicides with more than one ingredient or herbicides that are a key ingredient in mixtures. This table compares ingredients and allows users to quickly search for products that contain a particular ingredient. Each cell shows the amount of active ingredient at the low and high label rates in pounds of active ingredient or acid equivalent per acre.

Herbicides	2,4-D	mecoprop (MCPP)	dicamba	aminopyralid	carfentrazone	clopyralid	dichlorprop (2,4-DP)	fenoxaprop	flumioxazin	fluroxypyr	halauxifen-methyl	iodosulfuron	MCPA	mesotrione	metsulfuron	penoxsulam	pyraflufen ethyl	quinclorac[1]	sulfentrazone	thiencarbazone	triclopyr
2,4-D Amine 4, Barrage HF,[2] Clean Amine, Hardball, Saber, Shredder 2,4-D LV4, Shredder 2,4-D LV6, Shredder Amine 4, Shredder E-99, WEEDestroy AM40, Weedone LV4 EC, Weedar 64	1.0–1.9																				
2DQ	0.4–1.2		0.04–0.13															0.04–0.12			
3-D, Eliminate LO, EndRun, MEC Amine-D, Strike 3, Threesome, Trimec 899, Trimec 992, Trimec 1000, Triplet, Triplet Low Odor, Triplet SF, Vessel	0.5–1.2	0.15–0.33	0.05–0.11																		
4-Runner			0.09–0.16										0.6–1.0						0.025–0.042		0.10–0.16
4-Speed XT	0.5–1.1		0.06–0.14														0.001–0.003				0.06–0.14
Aquasweep, Chaser 2 Amine	0.7–1.4																				0.27–0.54
Avenue South	0.18–0.40		0.05–0.11													0.02–0.05			0.02–0.05		
Banvel, Clash, Cruise Control, Diablo, Sterling Blue, Vanquish, Vision			0.25–1.0																		
Battleship III										0.07–0.14			0.7–1.4								0.07–0.15
Brushmaster, Super Trimec	0.5–0.75		0.25–0.38				0.12–0.18														
Celsius WG			0.09–0.18									0.003–0.006								0.014–0.027	
Change Up			0.06–0.15							0.06–0.15			0.63–1.5								
Chaser, Crossroad, Everett	0.5–1.0																				0.25–0.5
Chaser Ultra 2							0.16–0.24			0.08–0.12			0.8–1.2								
Confront, 2-D						0.09–0.19															0.28–0.56
Cool Power, Three-Way Ester II			0.09–0.13										0.9–1.3								0.09–0.13
Dismiss, Spartan 4F[2]																			0.125–0.375		
Dismiss NXT, Spartan Charge[2]					0.01–0.04														0.13–0.38		
Drive XLR8, Quinclorac, QuinPro, Eject 75DF, others																		0.25–0.75			
Eliminate-D	0.25–0.37	0.17–0.25	0.06–0.10																		
Escalade 2	0.8–1.2		0.1–0.15							0.10–0.15											
Fahrenheit			0.06–0.25												0.01–0.04						
Foundation	0.8–1.0		0.08–0.10																0.024–0.030		0.20–0.25
GameOn	0.7–1.0									0.10–0.13	0.008–0.010										
Horsepower, Eliminate			0.10–0.14										1.0–1.4								0.10–0.14
Last Call			0.11–0.13					0.11–0.13		0.11–0.13											
Lockup, Sapphire																0.01–0.06					
LockUp Extra 2	1.5–1.8		0.12–0.14													0.015–0.017					

Active Ingredients in Commonly Used Postemergence Herbicide Combinations (*continued*)

Herbicides	2,4-D	mecoprop (MCPP)	dicamba	aminopyralid	carfentrazone	clopyralid	dichlorprop (2,4-DP)	fenoxaprop	flumioxazin	fluroxypyr	halauxifen-methyl	iodosulfuron	MCPA	mesotrione	metsulfuron	penoxsulam	pyraflufen ethyl	quinclorac[1]	sulfentrazone	thiencarbazone	triclopyr
Lontrel						0.09–0.5															
MCPA-4 Amine, MCPA Ester													0.5–1.85								
MCPP-p 4 Amine, Mecomec 2.5, Mecomec 4		1.0–2.0																			
Millennium Ultra 2	0.75–1.1		0.09–0.14			0.05–0.07															
Momentum 4-Score	0.5–0.6									0.11–0.15									0.03–0.04		0.11[illegible]–0.1[illegible]
Momentum FX2	0.8–1.1									0.10–0.14											0.1[illegible]–0.1[illegible]
Momentum Q, Quincept, Triad QC Select	0.9–1.0		0.11–0.13															0.66–0.75			
NativeKlean	0.5–0.9			0.06–0.11																	
Octane 2% SC																	0.001–0.006				
Patron 170	0.9–1.4						0.5–0.7														
Power Zone		0.11–0.33	0.06–0.17		0.01–0.03								0.6–1.7								
Q4 Plus	0.6–0.9		0.06–0.10															0.47–0.75	0.04–0.06		
Quicksilver T&O					0.01–0.1																
RedZone 2	0.5–1.1	0.13–0.29	0.05–0.11														0.001–0.003				
Solitaire, Solitaire WSL																		0.56–1.13	0.19–0.38		
Spartan Charge[2]					0.01–0.03														0.13–0.28		
SpeedZone	0.4–1.0	0.12–0.30	0.04–0.09		0.01–0.03																
SpeedZone Southern	0.3–0.8	0.05–0.15	0.01–0.04		0.01–0.03																
Spoiler	0.6–1.3	0.15–0.36					0.15–0.36														
Sublime			0.25											0.125							0.[illegible]
Sure Power, Tetra	0.5–0.9								0.005–0.009	0.06–0.10											0.0[illegible]–0.[illegible]
Surge, SureZone	0.5–0.7	0.17–0.25	0.08–0.12																0.02–0.03		
Tailspin										0.12–0.24											0.3[illegible]–0.[illegible]
Three-Way	0.6–1.2	0.33–0.65	0.06–0.11																		
Triad Select	0.5–1.2		0.05–0.11										0.14–0.31								
Triad SFZ Select	0.5–0.7		0.08–0.11										0.17–0.25						0.02–0.03		
Triad T Select	0.6–1.2		0.05–0.11										0.16–0.33								0.1[illegible]–0.[illegible]
Triamine	0.3–0.6	0.14–0.31					0.14–0.31														
Trimec Bentgrass Formula	0.15–0.23	0.24–0.36	0.06–0.09																		
Trimec Classic	0.4–1.0	0.12–0.27	0.05–0.11																		
Trimec Encore		0.14–0.31	0.07–0.15										0.7–1.6								
Trimec LAF-637	0.9–1.6	0.16–0.28	0.04–0.07																		
Trimec Southern	0.2–0.5	0.17–0.50	0.04–0.11																		
Tri-Power		0.14–0.32	0.07–0.15										0.7–1.5								
TruPower2	0.6–1.2	0.14–0.31	0.07–0.16																		
TruPower3	0.6–1.3	0.14–0.32	0.07–0.16																		
Turflon Ester Ultra, Triclopyr 4, Trycera																					0.5[illegible]
Turflon II Amine	0.9–1.2																				0.[illegible]–0.[illegible]
TZONE, TZONE SE, Triad TZ Select	0.4–0.9		0.05–0.10																0.02–0.03		0.[illegible]–0.[illegible]
Vista XRT, Flagstaff										0.13–0.50											

1 Quinclorac (Drive) is a grass herbicide active on crabgrass and foxtails with additional activity on certain broadleaf weeds like dandelion, speedwell, clover, bindweed, and black medic.
2 For use on sod farms only.

Sedge Control Herbicides

Sedge Control and Turfgrass Tolerance Ratings

	Sedge Control				Turf Tolerance								
	Sedges and Kyllinga				Cool-Season						Warm-Season		
Herbicide	annual sedge	false-green kyllinga	purple nutsedge	yellow nutsedge	annual bluegrass	creeping bentgrass	fine fescue	Kentucky bluegrass	perennial ryegrass	tall fescue	bermudagrass	buffalograss	zoysiagrass
2,4-D + dicamba + sulfentrazone + penoxsulam (Avenue South)	P	P	P	F	S	NR	S	S	S-I	S-I	S	S	S
2,4-D + fluroxypyr + triclopyr + sulfentrazone (Momentum 4-Score)	P	P	P	F	S	S	S	S	S	S	NR	NR	NR
2,4-D + MCPA + dicamba + sulfentrazone (Triad SFZ Select)	P	P	P	F	S	S	S	S	S	S	S	S	S
2,4-D + mecoprop + dicamba + sulfentrazone (Surge)	P	P	P	F	S	S	S	S	S	S	S	S	S
2,4-D + quinclorac + dicamba + sulfentrazone (Q4 Plus)	P	P	P	F	S	NR	S	S	S	S	S	NR	S
2,4-D + triclopyr + dicamba + sulfentrazone (Foundation)	P	P	P	F	S	NR	S	S	S	S	NR	NR	NR
2,4-D + triclopyr + dicamba + sulfentrazone (TZONE SE)	P	P	P	F	S	NR	S	S	S	S	S	NR	S
bentazon (Basagran T/O)	G	F	P	F	S	S	S	S	S	S	S	S	S
dimethenamid (Tower[1])	G	G	F	F-G	NR	NR	NR	NR	NR	NR	S	S	S
dimethenamid + pendimethalin (FreeHand)	G	G	F	F-G	NR	NR	NR	NR	NR	NR	S	S	S
flazasulfuron (Katana)	G	G	G-E	G-E	NR	NR	NR	NR	NR	NR	S	S	S
halosulfuron (SedgeHammer)	G	F	G	G-E	NR	S	S	S	S	S	S	S	S
halosulfuron + dicamba (Yukon[2])	G	F	G	G-E	NR	S	S	S	S	S	S	S-I	S
imazapic (Plateau)	F	F	F	F	NR	NR	NR	NR	NR	NR	S	S	NR
imazaquin (Image 70DG)	G	G-E	G	F	NR	NR	NR	NR	NR	NR	S	NR	S
imazosulfuron (Celero)	G	E	G-E	G-E	NR	S	S	S	S	S	S	NR	S
mesotrione (Tenacity)	P	P	P	G	NR	NR	S	S	S	S	NR	S	NR
metolachlor (Pennant MAGNUM)	G	F	F	G	NR	NR	NR	NR	NR	NR	S	NR	S
MSMA	G	F-G	P-F	F	S	I	I	S-I	S-I	S-I	S-I	I	S-I
pyrimisulfan (Arkon, Vexis)	G	F	F	F	NR	S	S	S	S	S	S	S	S
sulfentrazone (Dismiss, Spartan 4F[2])	G	F	P-F	G-E	NR	S-I	S-I	S	S	S-I	S	S	S[3]
sulfentrazone + carfentrazone (Dismiss NXT)	G	F	P-F	G-E	NR	S	S	S	S	S	S	S	S[3]
sulfentrazone + imazethapyr (Dismiss South)	G	F	E	G-E	NR	NR	NR	NR	NR	NR	S	S	S[3]
sulfentrazone + metsulfuron (Blindside)	G	F	P-F	G-E	NR	NR	NR	S	NR	S-I	S	S	S[3]
sulfentrazone + prodiamine (Echelon[4])	G	F	P-F	G-E	NR	S	S	S	S	S	S	S	S[3]
sulfentrazone + quinclorac (Solitare)	G	F	P-F	G-E	NR	NR	S-I	S	S	S	S	S	S[3]
sulfosulfuron (Certainty)	G	G-E	E	G-E	NR	NR[5]	NR	NR[5]	NR	NR	S	S	S
thiencarbazone + foramsulfuron + halosulfuron (Tribute TOTAL)	G	F	G	G-E	NR	NR	NR	NR	NR	NR	S	NR	S
thiencarbazone + iodosulfuron + halosulfuron (Celsius XTRA)	G	F	G	G-E	NR	NR	NR	NR	NR	NR	S	S	S
trifloxysulfuron (Monument, Recognition)	G	E	E	G	NR	NR	NR	NR	NR	NR	S	S-I	S

Rating Key: E = excellent (≥90% control). G = good (75%-90% control). F = fair (50%-75% control). P = poor control (≤50% control). Blank = no data. S = safe at labeled rates on healthy, mature turf. I = intermediate safety or some injury may occur; may cause minor damage to mature, healthy turf. Consider using the lower end of the rate range. Do not apply to turf under stress. NR = not registered for use on this species.

These are relative control rankings based on research of the efficacy of professional products. Weed control will vary with environmental conditions, application timing, turfgrass vigor, and other factors. This table is intended only as a guide. Repeat applications are needed for complete control of sedges with all herbicides.

1 For use on cool- and warm-season grasses on golf courses (see label) and warm-season use only on sod farms, commercial and recreational turf, and residential turf.

2 For use on sod farms only.

3 Safety varies by zoysiagrass cultivar.

4 Labeled for preemergence control of sedges in addition to postemergence control.

5 A label change in 2011 no longer allows for the use of Certainty turf herbicide on cool-season turf. These changes are effective on product packaged and shipped after May 2011. All cool-season turfgrass uses for Certainty have been removed from the 2011-2012 label (product packaged and shipped after May 2011). Product packaged before these changes can continue to be used according to label directions in cool-season turf. Previous label stipulations recommended that Certainty was safe for use on creeping bentgrass, 100% Kentucky bluegrass turf, or for use on Kentucky bluegrass + perennial ryegrass and/or fine fescue mixtures, but not safe for use on tall fescue + Kentucky bluegrass mixtures.

Preemergence and Postemergence Sedge Control Herbicides

Refer to herbicide labels for weeds controlled and specific turfgrass species use instructions.

Trade Name (product/A)	Common Name (lb ai/A)	Weeds Controlled	Turfgrasses	Comments
Avenue South (5–8 pt) L	penoxsulam + sulfentrazone + 2,4-D + dicamba (see label)	suppresses yellow nutsedge; controls broadleaf weeds	annual bluegrass, bermudagrass, buffalograss, fine fescue, Kentucky bluegrass, perennial ryegrass, tall fescue, zoysiagrass	Do not use on creeping bentgrass. Do not apply to golf course putting greens, tees, or collars. Do not exceed 2 applications per year. Avoid applications when warm-season grasses are transitioning into or out of dormancy. Perennial ryegrass and tall fescue are less tolerant than other turfgrasses. See label for more information.
Arkon (52–74 fl oz) Vexis G (109–240 lb) NY	pyrimisulfan (0.027–0.06)	suppresses and controls kyllinga, purple nutsedge, yellow nutsedge, and some broadleaf weeds	bermudagrass, buffalograss, creeping bentgrass, fine fescue, Kentucky bluegrass, perennial ryegrass, tall fescue, zoysiagrass	Apply to emerged sedge actively growing. Apply granules to wet or dry turf and water-in granules within 48 hours for best results. Do not collect grass clippings from treated areas to use as mulch around plants. Do not apply Vexis to golf course putting greens or creeping bentgrass <0.5 inch tall. Arkon is safe to use on short-cut creeping bentgrass including putting greens.
Basagran T/O (2 pt) LESCOGran (2 pt)	bentazon (1.0)	suppresses and controls yellow nutsedge, annual sedge, kyllinga, many broadleaf annual weeds, and Canada thistle	annual bluegrass, bermudagrass, buffalograss, creeping bentgrass, fine fescue, Kentucky bluegrass, perennial ryegrass, tall fescue, zoysiagrass	Apply to emerged sedge actively growing under good soil moisture conditions. Spray with at least 1 gal/1,000 ft^2 of water volume and mix with COC or MSO. Apply 2 pt/A, and follow 10–14 days later with an additional application if necessary. Thorough spray coverage is essential for acceptable control. Spot spraying may result in turf injury. Do not use on golf course putting greens.
Blindside (3.25–10.0 oz) W L NY	sulfentrazone + metsulfuron (see label)	suppresses and controls sedges and broadleaf weeds	bermudagrass, Kentucky bluegrass, tall fescue, zoysiagrass	**Use caution when using this product on cool-season grasses,** especially tall fescue, which is more sensitive to this herbicide than Kentucky bluegrass. Controls a wide range of broadleaf weeds, sedges, and wild garlic/onion. Do not add a surfactant unless previous experience indicates this will not increase turf injury.
Celero (8–14 oz) NY	imazosulfuron (0.38–0.66)	suppresses and controls false-green kyllinga, purple nutsedge, and yellow nutsedge; also controls various broadleaf weeds	bermudagrass, creeping bentgrass, fine fescue, Kentucky bluegrass, perennial ryegrass, tall fescue, zoysiagrass	Add NIS at 0.25% (v/v). More than 1 application may be needed for complete weed control. Repeat application 21 days after the initial application if needed. Do not apply to moist or wet turf.
Celsius XTRA (5–10 oz/A) W	thiencarbazone + iodosulfuron + halosulfuron (see label)	broadleaf weeds, some annual and perennial grasses, and sedges and kyllinga	bermudagrass, buffalograss, zoysiagrass	**Do not use on cool-season turf.** Do not apply where runoff onto cool-season grasses may occur. Add NIS at 0.25% to 0.5% (v/v) unless air temperature exceeds 90°F coupled with high humidity at the time of application. Do not use more than 15 oz/A per year.
Certainty (1.25–2.0 oz) W	sulfosulfuron (0.058–0.092)	suppresses and controls annual sedge, kyllinga, purple nutsedge, and yellow nutsedge; also controls annual bluegrass, quackgrass, ryegrasses, tall fescue, and johnsongrass	bermudagrass, zoysiagrass	**Do not use on cool-season turf.** Add NIS at 0.25% (v/v). May be used on commercial and residential turf. Make a second application after 30 days if needed. A label change in 2011 no longer allows for use on cool-season turf. Product packaged before these changes can be used according to directions on the affixed label.
Dismiss (0.25–0.75 pt) Spartan 4F (*sod farms only*, 0.25–0.75 pt) NY	sulfentrazone (0.125–0.375)	suppresses and controls annual sedge, kyllinga, purple nutsedge, and yellow nutsedge; also controls various broadleaf weeds	bermudagrass, buffalograss, creeping bentgrass, fine fescue, Kentucky bluegrass, perennial ryegrass, tall fescue, zoysiagrass	Adding a surfactant is not recommended. Surfactants may cause temporary discoloration. Application rate varies by turf species. Good coverage is needed for optimum control. Rates < 12 fl. oz/A will generally suppress most sedges for at least 60 days but will require a second application.

W This product is labeled for use primarily in warm-season turf. Read the label for a list of tolerant grasses.
NY Use or sale of this product has restrictions in New York State or Long Island, New York (Nassau and Suffolk counties). Refer to the label for more information.
L This product is only registered in specific North Central states. Check with local contacts for availability.

Preemergence and Postemergence Sedge Control Herbicides (*continued*)

Trade Name (product/A)	Common Name (lb ai/A)	Weeds Controlled	Turfgrasses	Comments
Dismiss NXT (5.1–15.25 fl oz) Spartan Charge (*sod farms only,* 5.1–11.5 fl oz) NY	carfentrazone + sulfentrazone (see label)	suppresses and controls annual sedge, kyllinga, purple nutsedge, and yellow nutsedge; also controls sandbur and various broadleaf weeds	bermudagrass, buffalograss, creeping bentgrass, fine fescue, Kentucky bluegrass, perennial ryegrass, tall fescue, zoysiagrass	Adding a surfactant is not recommended. Surfactants may cause temporary discoloration. Good coverage is needed for optimum control. A second application may be needed for best control of sedges.
Dismiss South (4.0 to 14.4 fl oz) W L NY	sulfentrazone + imazethapyr (see label)	suppresses and controls annual sedge, kyllinga, purple nutsedge, and yellow nutsedge; also controls dallisgrass, sandbur, and various broadleaf weeds	bermudagrass, buffalograss, zoysiagrass	**Do not use on cool-season turf.** Do not apply to golf course putting greens or tees. Adding a surfactant is not recommended. Surfactants may cause temporary discoloration. Do not apply Dismiss South < 7 days after or 7 days before a trinexapac-ethyl application.
Echelon 4SC (8–36 fl oz) Echelon G available on fertilizer carrier (see label) NY	sulfentrazone + prodiamine (see label)	suppresses and controls annual sedge, kyllinga, purple nutsedge, and yellow nutsedge; also controls various annual grassy weeds	bermudagrass, buffalograss, creeping bentgrass, fine fescue, Kentucky bluegrass, perennial ryegrass, tall fescue, zoysiagrass	Provides preemergence control of sedges. Apply only to well-established turf. Do not apply to golf course putting greens or tees. Do not exceed 36 oz/A per year. Wait 3–6 months before reseeding after application.
Foundation (3.25–4.0 pt) NY	2,4-D + triclopyr + dicamba + sulfentrazone (see label)	yellow nutsedge suppression; also controls various broadleaf weeds	annual bluegrass, Kentucky bluegrass, perennial ryegrass, fine fescue, tall fescue	Controls a broader spectrum of weeds than 2,4-D alone. Not for use on putting greens or tees. Do not apply when temperatures are > 90°F. Provides yellow nutsedge suppression only. For newly seeded areas, delay application until after second mowing. Do not use adjuvants.
FreeHand 1.75G (100–200 lb) W NY	dimethenamid + pendimethalin (see label)	suppresses and controls annual sedge, kyllinga, purple nutsedge, yellow nutsedge, and various grassy weeds and broadleaves	established bermudagrass, buffalograss, zoysiagrass	**Do not use on cool-season turf.** Apply to well-established turf. Delay seeding for at least 3 months following the last application. Apply to warm-season turfgrass following spring transition when soil temperature is 55°F or warmer. Rainfall or irrigation (0.5 inch) after application will improve control.
Image 70 DG (8.6–11.4 oz) W L	imazaquin (0.37–0.5)	suppresses and controls annual sedge, kyllinga, purple nutsedge, and yellow nutsedge; controls perennial ryegrass, tall fescue, and certain broadleaves	bermudagrass, zoysiagrass	**Do not use on cool-season turf.** Add NIS (0.25% = 1.0 qt/100 gal). Do not apply to golf course putting greens. Use only on well-established turf. Do not apply when turf is emerging from winter dormancy.
Katana (1.5–3.0 oz) W NY	flazasulfuron (0.024–0.047)	suppresses and controls annual sedge, kyllinga, purple nutsedge, and yellow nutsedge; also controls annual bluegrass, ryegrasses, and tall fescue	bermudagrass, buffalograss, zoysiagrass	**Do not use on cool-season turf.** Add NIS (0.25% = 1.0 qt/100 gal). Irrigate lightly after application. Use with caution in warm-season turf bordering cool-season turf because it can track from one area to another. A 25-foot border area around cool-season grasses is recommended.
Momentum 4-Score (3.5–4.5 pt) NY	2,4-D + fluroxypyr + triclopyr + sulfentrazone (see label)	yellow nutsedge suppression; also controls various broadleaf weeds	annual bluegrass, creeping bentgrass, Kentucky bluegrass, perennial ryegrass, fine fescue, tall fescue	Controls a broader spectrum of weeds than 2,4-D alone. Not for use on putting greens or tees. Do not apply when temperatures are > 90°F. Provides yellow nutsedge suppression only. For newly seeded areas, delay application until after second mowing. Do not use adjuvants.
Monument 75 WG (0.35–0.53 oz) W L NY	trifloxysulfuron (0.016–0.025)	suppresses and controls annual sedge, kyllinga, purple nutsedge, and yellow nutsedge; controls annual bluegrass, ryegrasses, tall fescue, and certain broadleaves.	bermudagrass, buffalograss, zoysiagrass	**Do not use on cool-season turf.** Add NIS at 0.25% (v/v). Do not apply where runoff onto cool-season grasses may occur. Repeat application when regrowth appears. To reduce potential movement, water-in lightly 2–3 hours after application to remove product from turf foliage. Allow turf to dry before allowing traffic onto treated areas.

W This product is labeled for use primarily in warm-season turf. Read the label for a list of tolerant grasses.

NY Use or sale of this product has restrictions in New York State or Long Island, New York (Nassau and Suffolk counties). Refer to the label for more information.

L This product is only registered in specific North Central states. Check with local contacts for availability.

Preemergence and Postemergence Sedge Control Herbicides (*continued*)

Trade Name (product/A)	Common Name (lb ai/A)	Weeds Controlled	Turfgrasses	Comments
MSMA 6 Plus (2.7 pt) MSMA 6.6 (2.5–5.0 pt) Target 6 Plus (2.7 pt) Target 6.6 (2.7 pt) others L	MSMA (2.0–2.2)	suppresses and controls annual sedge, kyllinga, purple nutsedge, yellow nutsedge, crabgrass, dallisgrass, sandbur, and some broadleaf weeds	annual bluegrass, bermudagrass, Kentucky bluegrass, zoysiagrass	Temporary discoloration of turf will occur. Sedges may require multiple applications. **MSMA use is being phased out by the EPA**—see *Can I still use MSMA?*, page 66 for details. Sale prohibited to residential turf after Dec. 31, 2010. Existing supplies can be used on lawns per label instructions. Starting in 2011, only spot treatments (100 ft² maximum per spot) are allowed on existing golf courses. Sod farms and golf courses only.
Pennant MAGNUM (1.3–2.6 pt) W NY	metolachlor (1.24–2.48)	suppresses and controls annual sedge, annual kyllinga, yellow nutsedge, annual grasses, and annual broadleaf weeds	bermudagrass, zoysiagrass	**Do not use on cool-season turf.** Provides preemergence sedge control. Apply before weeds emerge and water-in with 0.5 inch of irrigation if rainfall does not occur within 7 days. Delayed spring green-up, temporary slowing of growth, and yellowing may occur following application. Do not apply to newly seeded areas. Do not reseed within 4 months of application.
Plateau (2 to 6 fl oz) W NY	imazapic (0.03–0.09)	yellow and purple nutsedge	bermudagrass, buffalograss	**Do not use on cool-season turf.** Tolerance to this herbicide varies by cultivar. Rates of 4–6 fl oz/A will suppress both nutsedge and bermudagrass seedheads. Higher rates may injure bermudagrass.
Q4 Plus (5–8 pt) NY	quinclorac + sulfentrazone + 2,4-D + dicamba (see label)	suppresses annual sedge, kyllinga, purple nutsedge, and yellow nutsedge; also controls broadleaf weeds and some annual grasses	annual bluegrass, bermudagrass, fine fescue, Kentucky bluegrass, perennial ryegrass, tall fescue, zoysiagrass	Do not use on creeping bentgrass. Do not apply to golf course putting greens, tees, or collars. Limit of 2 applications per year. Avoid applications when warm-season grasses are transitioning into or out of dormancy.
Recognition (0.37–1.95 oz/A)	trifloxysulfuron + metcamifen	suppresses and controls sedges, annual bluegrass, ryegrasses, tall fescue, and certain broadleaf weeds	bermudagrass, buffalograss, zoysiagrass	**Do not use on cool-season turf.** Add NIS at 0.25% (v/v). Do not apply where runoff onto cool-season grasses may occur. Repeat application when regrowth appears. To reduce potential movement, water-in lightly 2–3 hours after application to remove product from turf foliage. Allow turf to dry before allowing traffic onto treated areas. Tank-mix with other herbicides to improve the spectrum of control. A 0.53 oz/A rate of Monument is equivalent trifloxysulfuron to a 1.95 oz/A rate of Recognition. Metcamifen is an herbicide safener and is not listed on the product label as an active ingredient but is listed as an ingredient on the SDS.
SedgeHammer (0.66–1.33 oz) ProSedge (0.66–1.33 oz) Sandea (*sod farms only*, 0.66–1.33 oz) Profine75 (*sod farms only*, 0.66–1.33 oz) SedgeHammer+ (21 oz)	halosulfuron (0.031–0.068)	suppresses and controls annual sedge, kyllinga, purple nutsedge, and yellow nutsedge	annual bluegrass, bermudagrass, buffalograss, creeping bentgrass, fine fescue, Kentucky bluegrass, perennial ryegrass, tall fescue, zoysiagrass	The SedgeHammer+ formulation includes surfactant, so adding it isn't necessary. Use NIS at 0.25–0.5% (v/v) with the 75DF formulation. Can make a second application 6 weeks after the first. Do not exceed 0.125 lb/ai/A in a single season. Not labeled for putting greens. For spot treatment, mix 0.9 gram of 75DF formulated halosulfuron in 1 gal of water with 0.33 oz (2 tsp) surfactant. For spot treatment with SedgeHammer+, use 0.5 oz/1,000 ft².

W This product is labeled for use primarily in warm-season turf. Read the label for a list of tolerant grasses.
NY Use or sale of this product has restrictions in New York State or Long Island, New York (Nassau and Suffolk counties). Refer to the label for more information.
L This product is registered only in specific North Central states. Check with local contacts for availability.

Preemergence and Postemergence Sedge Control Herbicides (*continued*)

Trade Name (product/A)	Common Name (lb ai/A)	Weeds Controlled	Turfgrasses	Comments
Solitare (1-2 lb) Solitaire WSL (5.4–10.7 qt) NY	sulfentrazone + quinclorac (0.75–1.5)	suppresses and controls annual sedge, kyllinga, purple nutsedge, and yellow nutsedge; also controls various broadleaf weeds	bermudagrass, buffalograss, fine fescue, Kentucky bluegrass, perennial ryegrass, tall fescue, zoysiagrass	Do not apply to golf course putting greens, collars, or tees. Do not apply directly to landscape ornamentals or to ornamental beds. Do not apply with surfactants unless compatibility tests have previously demonstrated compatibility and safety on grass type. Seeding should be delayed 4 weeks after application. Do not use on newly seeded turfgrasses until they have been mowed at least twice. Solitaire WSL is labeled for use in fine fescues.
Solitare (1-2 lb) Solitaire WSL (5.4–10.7 qt) NY	sulfentrazone + quinclorac (0.75–1.5)	suppresses and controls annual sedge, kyllinga, purple nutsedge, and yellow nutsedge; also controls various broadleaf weeds	bermudagrass, buffalograss, fine fescue, Kentucky bluegrass, perennial ryegrass, tall fescue, zoysiagrass	Do not apply to golf course putting greens, collars, or tees. Do not apply directly to landscape ornamentals or to ornamental beds. Do not apply with surfactants unless compatibility tests have previously demonstrated compatibility and safety on grass type. Seeding should be delayed 4 weeks after application. Do not use on newly seeded turfgrasses until they have been mowed at least twice. Solitaire WSL is labeled for use in fine fescues.
Surge (2.75–4.0 pt) SureZone (2.75–4.5 pt) NY	sulfentrazone + 2,4-D + MCPP + dicamba (see label)	yellow nutsedge suppression; also controls various broadleaf weeds	annual bluegrass, bermudagrass, buffalograss, creeping bentgrass, fine fescue, Kentucky bluegrass, perennial ryegrass, tall fescue, zoysiagrass	Do not apply when temperatures are > 90°F or during spring transition. Provides yellow nutsedge suppression only. For newly seeded areas, delay application until after the second mowing.
Tenacity (5–8 fl oz)	mesotrione (0.156–0.25)	yellow nutsedge control; also controls various broadleaf weeds and grasses	buffalograss, fine fescue, Kentucky bluegrass, perennial ryegrass, tall fescue	Causes a bleaching (whitening) effect to susceptible species. Do not exceed 16 fl oz/A per year. Add NIS at 0.25% (v/v). Make second application 2 weeks after the first.
Tower (21–32 fl oz) W NY	dimethenamid (1.0–1.5)	suppresses and controls annual sedge, kyllinga, purple nutsedge, and yellow nutsedge, and various grassy weeds and broadleaves	bermudagrass, buffalograss, zoysiagrass	On golf courses: can be used on cool- and warm-season turf. **For all other turf areas: warm-season turf only**. Tower provides some preemergence control of sedges. Apply to well-established turf or new seedlings that have been mown at least 4 times. Do not apply more frequently than every 35 days. Do not exceed 64 fl oz/A (3.0 lb ai) of per year.
Triad SFZ Select (2.75–4.0 pt/A) NY	2,4-D + MCPA + dicamba + sulfentrazone (see label)	yellow nutsedge suppression; also controls various broadleaf weeds	annual bluegrass, bermudagrass, buffalograss, creeping bentgrass, fine fescues, Kentucky bluegrass, perennial ryegrass, tall fescue, zoysiagrass	Controls a broader spectrum of weeds than 2,4-D alone. Not for use on putting greens or tees. Avoid applications when temperatures are > 90°F. Delay reseeding 3 weeks after treatment. Do not apply to newly seeded grasses until mown twice. Limit of 2 applications per year. Do not use adjuvants.
Tribute TOTAL (1.0–3.2 oz) W NY	thiencarbazone + foramsulfuron + halosulfuron (see label)	broadleaf weeds, annual and perennial grassy weeds, and sedges and kyllinga	bermudagrass, zoysiagrass	**Do not use on cool-season turf.** Do not apply where runoff onto cool-season grasses may occur. Add NIS at 0.25%–0.5% (v/v). Repeat application when regrowth appears. See label for instructions on using Tribute TOTAL next to sensitive areas such as golf course putting greens. Allow turf to dry before allowing traffic onto treated areas.
TZONE SE (2–4 pt) TZONE (2–4 pt) NY	triclopyr + sulfentrazone + 2,4-D + dicamba (see label)	yellow nutsedge suppression, also controls various broadleaf weeds	annual bluegrass, bermudagrass, fine fescue, Kentucky bluegrass, perennial ryegrass, tall fescue, zoysiagrass	Rate varies by turf species. Use 2–2.25 pt/A on bermudagrass and zoysiagrass, and 3.25–4 pt/A on cool-season turf. Limit of 2 applications per year. Delay application to newly seeded turf until after the second or third mowing. Delay application until 3–4 weeks after sodding, sprigging, or plugging turf. Treated areas may be reseeded 3 weeks after application. Do not use when temperatures exceed 85°F. Do not spray when warm-season turfgrass is emerging from winter dormancy.
Yukon (*sod farms only*, 4–8 oz)	halosulfuron + dicamba (see label)	suppresses and controls annual sedge, kyllinga, purple nutsedge, and yellow nutsedge; also controls various broadleaf weeds	annual bluegrass, bermudagrass, creeping bentgrass, fine fescue, Kentucky bluegrass, perennial ryegrass, tall fescue, zoysiagrass	Use NIS at 0.25%–0.5% (v/v). A second application may be made 6 weeks after the first. Do not exceed 8 oz/A per season.

W This product is labeled for use primarily in warm-season turf. Read the label for a list of tolerant grasses.
NY Use or sale of this product has restrictions in New York State or Long Island, New York (Nassau and Suffolk counties). Refer to the label for more information.

Distribution of Common Sedges in Managed Turfgrass

These maps show the distribution areas of various sedge species found in the United States.

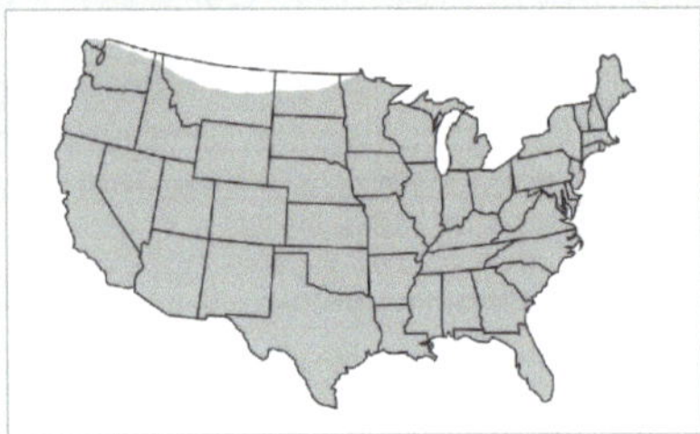

Yellow nutsedge (*Cyperus esculentus*)

Purple nutsedge (*Cyperus rotundus*)

Annual sedge (*Cyperus compressus*)

Shining flat sedge (*Cyperus bipartitus*)

Green kyllinga (*Kyllinga brevifolia*)

False-green kyllinga (*Kyllinga gracillima*)

Comparisons of Sulfentrazone Rates in Various Herbicides

			Ingredient												
Trade Name	**Rate of Product**	**Rate of Sulfentrazone[1] (lb ai/A)**	**2,4-D**	**carfentrazone**	**dicamba**	**fluroxypyr**	**imazethapyr**	**MCPA**	**MCPP**	**metsulfuron**	**penoxsulam**	**prodiamine**	**quinclorac**	**sulfentrazone**	**triclopyr**
4-Runner	2.75–4.5 pt/A	0.025–0.042			■			■						■	■
Avenue South	2.7–6.0 pt/A	0.020–0.046	■		■						■			■	
Blindside[2]	3.25–10 oz/A	0.122–0.375								■				■	
Dismiss, Spartan 4F[3]	4–12 fl oz/A[4]	0.125–0.375												■	
Dismiss NXT	5.1–15.25 fl oz/A[5]	0.125–0.375		■										■	
Dismiss South	9.5–14.4 fl oz/A	0.250–0.375					■							■	
Echelon 4SC	8–36 fl oz/A[6]	0.083–0.375										■		■	
Foundation	3.25–4.0 pt/A	0.024–0.030	■		■									■	■
Momentum 4-Score	3.5–4.5 pt/A	0.034–0.044	■			■								■	■
Q4 Plus	5.0–8.0 pt/A	0.040–0.060	■		■								■	■	
Solitaire, Solitaire WSL	16–32 oz/A	0.187–0.375											■	■	
Spartan Charge[3]	5.1–11.5 fl oz/A[7]	0.125–0.280		■										■	
Surge, SureZone	2.75–4.0 pt/A	0.021–0.030	■		■									■	
Triad SFZ Select	2.75–4.0 pt/A	0.020–0.030	■		■			■						■	
TZONE, TZONE SE	2.0–4.0 pt/A	0.015–0.030	■		■									■	■

1 Approximately 0.1875 lb/A of sulfentrazone is needed for excellent (≥90%) yellow nutsedge control. Lower application rates can provide fair (50%–75%) control to good (75%–90%) control.
2 The Blindside label recommends use rates of 3.25–5.0 oz/A for Kentucky bluegrass and tall fescue. At least 5 oz/A (0.1875 lb /A of sulfentrazone) is needed for acceptable yellow nutsedge control. Injury can occur on tall fescue at the 5 oz/A rate. Bermudagrass and zoysiagrass can be treated at 5–10 oz/A.
3 For use on sod farms only.
4 The Dismiss and Spartan 4F labels list rates as low as 4 fl oz/A. However, 6 fl oz/A (0.1875 lb/A of sulfentrazone) is needed for acceptable yellow nutsedge control. All species of turf on the label may be treated at 6 fl oz/A except creeping bentgrass.
5 The Dismiss NXT label lists rates as low as 5.1 fl oz/A. However, a rate of 7.6 fl oz/A (0.1875 lb/A of sulfentrazone) is necessary for acceptable yellow nutsedge control. All turf species on the label may be treated at ≥7.6 fl oz/A except creeping bentgrass.
6 The Echelon 4SC label recommends use rates of 8–12 fl oz/A for creeping bentgrass, fine fescue, and perennial ryegrass. However, at least 18 fl oz/A (0.1875 lb/A of sulfentrazone) is needed for acceptable yellow nutsedge control. Kentucky bluegrass, tall fescue, zoysiagrass, and bermudagrass can be treated with this higher rate of Echelon. Only bermudagrass can be treated at the 36 fl oz/A rate.
7 The Spartan Charge label recommends use rates as low as 5.1 fl oz/A for cool-season turf. Rates of at least 7.6 fl oz/A (0.1875 lb/A of sulfentrazone) will be required for acceptable yellow nutsedge control. All species of turf on the label may be treated at 7.6 fl oz/A except creeping bentgrass.

Putting Green Herbicides

Preemergence Weed Control in Creeping Bentgrass Putting Greens

Refer to herbicide labels for weeds controlled and specific turfgrass species use instructions.

Trade Name (product/Acre)	Common Name (lb ai or ae/A)	Weeds Controlled	Comments
Bensumec 4LF (1.88–3.13 gal) Pre-San 12.5G (80–100 lb) Weedgrass Preventer (146 lb)	bensulide (see label)	controls annual grasses and selected broadleaf weeds	Apply in spring for preemergence control of annual grassy weeds such as crabgrass and goosegrass. Apply high rate in late summer for annual bluegrass control. Irrigate immediately after treatment. Delay reseeding for 4 months after treatment. Limit of 2 applications per year on putting greens. Do not use on a putting green composed of more than 50% annual bluegrass (*Poa annua*).
Goosegrass/ Crabgrass Control (114 lb)	bensulide + oxadiazon (6.0 + 1.5)	controls summer annual grasses	To prevent injury, apply an initial treatment to dry foliage at half the rate (57 lb/A) followed by a treatment 10–14 days later at the half rate. Lightly irrigate after treatment. Delay reseeding 5 months after treatment. See label for precautions about use on putting greens. Limit of 2 applications per year on putting greens.
Fertilizers with 0.164% dithiopyr (152 lb) NY	dithiopyr (0.25)	controls summer annual grasses	Various products from Andersons Golf Products. Use of this product on certain older varieties of bentgrass (such as Cohansey, Carmen, Seaside, and Washington) may result in undesirable injury. Not all creeping bentgrass varieties have been tested. Avoid applying to stressed putting greens. Undesirable thinning or yellowing of *Poa annua* may occur during stressful conditions when using this product. Using this product on creeping bentgrass that does not have good root development may result in injury. Application rate is 0.25 lb ai/A or 152 lb/A using 0.164% dithiopyr formulation.
PoaCure (0.2–0.4 gal) NY	methiozolin (0.46–0.92)	Preemergence and postemergence control of annual bluegrass; preemergence crabgrass and goosegrass control	For use on golf course putting greens, fairways, and other cool- and warm-season turf areas. Water -n sprays with 0.1 to 0.2-inch of irrigation after application. The herbicide is slow acting, taking 2 to 4 weeks to control weeds. Do not treat putting greens under stress or within 1 week of aerification. Check the label for application timing, rate, and frequency recommendations.

NY Use or sale of this product has restrictions in New York State or Long Island, New York (Nassau and Suffolk counties). Refer to the label for more information.

Tips for Herbicide Use on Golf Course Putting Greens

Although most golf course superintendents have few weed problems (other than annual bluegrass in their putting greens), weeds do occasionally invade because of the stress caused by ball marks, disease, close mowing, and traffic. If only a handful of weeds are present in a single green, the most efficient control method is to mechanically remove the weeds with a pocketknife. Herbicide applications are only warranted when weed pressure is significant. In these cases, Purdue University and University of Tennessee research has confirmed that herbicides labeled for use on putting greens can be used safely and successfully to control weeds when applicators follow label directions. Specifically, our findings were that broadleaf herbicides labeled for putting green use are safe to use at labeled rates in the spring and fall (with fall applications being less likely to injure the turf). Unacceptable injury can occur if using higher-than-labeled herbicide rates (such as from spot applications).

More herbicides are registered for use on golf course fairways and tees than on putting greens. Considering the scope of efficacy and tolerance testing required across a wide geographic region before an herbicide receives labeling, there is likely good reason a particular product is not labeled for putting green use: it could result in undesirable injury. Further, companies do not register herbicides for use on putting greens due to the liability of potentially causing herbicide injury and because putting greens make up such a small percentage of total turf acreage. Some herbicide labels are silent on putting green use and neither explicitly restrict nor allow it. This places all liability on the end user, so use caution with these products. For these reasons, only use products with specific label instructions for putting green use.

Postemergence Weed Control in Putting Greens

Refer to herbicide labels for weeds controlled and specific turfgrass species use instructions.

Trade Name (product/A)	Common Name (lb ai/A)	Weeds Controlled	Turfgrasses	Comments
2DQ (1–3 pt) NY	2,4-D + dicamba + quinclorac (see label)	broadleaf weeds	creeping bentgrass	Controls a broader spectrum of weeds than 2,4-D alone. Do not apply this product when temperatures are > 90°F. Limit of 2 applications per year.
4-Speed XT (1.8 pt)	2,4-D + triclopyr + dicamba + pyraflufen ethyl (see label)	broadleaf weeds	annual bluegrass, creeping bentgrass	Controls a broader spectrum of broadleaf weeds than 2,4-D alone. Limit of 2 applications per site per year. Avoid applications during periods when turf is under stress due to high heat, humidity, or reduced moisture. Slight turf yellowing will disappear after about 1 week.
Arkon (52–74 fl oz)	pyrimisulfan (0.04–0.06)	sedges and kyllinga	creeping bentgrass	Safe for use on all major cool- and warm-season turfgrass species. Weed control may not be evident until 2 to 3 weeks after application. While annual bluegrass is not listed on the label, research indicates the product is safe to apply to it.
Banvel (≤1 pt)) Diablo (≤1 pt))	dicamba (≤0.5)	broadleaf weeds	annual bluegrass, creeping bentgrass	**Restricted use pesticide in some states because of crop injury concerns.** Label neither allows nor restricts applications to putting greens. Use (≤1 pt)/A. As little as 4 fl oz/A will provide adequate weed control.
Mecomec 2.5 (4 pt) Mecomec 4 (2.5 pt)	mecoprop-p (MCPP) (1.0–1.38)	broadleaf weeds	annual bluegrass, creeping bentgrass	Use only on actively growing turf that is not under stress. Do not apply to bentgrass in the heat of summer.
PoaCure (0.2–0.4 gal) NY	methiozolin (0.46–0.92)	preemergence and postemergence control of annual bluegrass; postemergence rough bluegrass contro.	creeping bentgrass	For use on golf course putting greens, fairways, and other cool- and warm-season turf areas. Water in sprays with 0.1 to 0.2-inch of irrigation after application. The herbicide is slow acting, taking 2 to 4 weeks to control weeds. Do not treat putting greens under stress or within 1 week of aerification. Check the label for application timing, rate, and frequency recommendations.
Quicksilver T&O (1.0–6.7 oz)	carfentrazone (0.015–0.1)	silvery thread moss	annual bluegrass, creeping bentgrass	Provides excellent moss control when applied at 2.0–6.7 oz/A when temperatures are < 85°F. Apply as often as every 2 weeks to putting greens infested with silvery thread moss. Annual bluegrass can be damaged at rates > 2.0 oz/A. Use NIS at 0.25% (v/v). Do not apply to bentgrass when temperatures exceed 90°F. Do not apply if Bensumec (bensulide) has been applied within the previous 75 days.
RedZone 2 (1.8 pt)	2,4-D + MCPP + dicamba + pyraflufen ethyl (see label)	broadleaf weeds	annual bluegrass, creeping bentgrass	Controls a broader spectrum of broadleaf weeds than 2,4-D alone. Limit of 2 applications per site per year. Avoid applications during periods when turf is under stress. Slight turf yellowing will disappear after about 1 week.
Triad Select (1.8 pt/A)	2,4-D + MCPA +dicamba (see label)	broadleaf weeds	creeping bentgrass	Do not exceed 0.67 fl oz/1,000 ft² on creeping bentgrass putting greens using a spray volume of 3.3 gal/1,000 ft². Do not apply when temperatures exceed 90°F.
Trimec Bentgrass (2.7 pt)	2,4-D + mecoprop + dicamba (see label)	broadleaf weeds	creeping bentgrass	Contains 4.5 times less 2,4-D than Trimec Classic, providing for better safety on creeping bentgrass. Do not apply to bentgrass under stress. Do not apply when temperatures exceed 85°F.
Eliminate LO, Threesome, Trimec Classic, Trimec 899, Triplet, Triplet Low Odor, Triplet SF, TruPower2, TruPower3 (1.8 pt)	2,4-D + mecoprop + dicamba (see label)	broadleaf weeds	annual bluegrass, creeping bentgrass	Do not exceed 1.0 fl oz/1,000 ft² on creeping bentgrass putting greens using a spray volume of 5 gal/1,000 ft². Do not apply to bentgrass under stress. Do not apply when temperatures exceed 85°F (some labels state 90°F).
Trimec Encore (1.8 pt) Tri-Power (1.8 pt)	MCPA + MCPP + dicamba (see label)	broadleaf weeds	annual bluegrass, creeping bentgrass	Do not exceed 1.0 fl oz/1,500 ft² on creeping bentgrass putting greens using a spray volume of 5 gal/1,000 ft². Do not apply to bentgrass under stress. Do not apply when temperatures exceed 85°F. Slight yellowing will occur within 1 week.

NY Use or sale of this product has restrictions in New York State or Long Island, New York (Nassau and Suffolk counties). Refer to the label for more information.

Common and Trade Names and Mechanism of Action of Registered Herbicides

Sorted by common name.

Common Name	Trade Name(s)	Mechanism of Action[1]
2,4-D	2,4-D Amine 4, 2,4-D LV 6, Amine 400, Barrage HF, Base Camp Amine 4, Clean Amine, De-Amine 4, Hardball, Hi-Dep, LV 400, Opti-Amine, Pasture Pro, Saber, Shredder 2,4-D LV4, Shredder 2,4-D LV6, Shredder Amine 4, Shredder E-99, Solution, WEEDestroy AM40, Weedone LV4 EC, Weedar 64	4
2,4-D + aminopyralid	NativeKlean	4 + 4
2,4-D + dicamba	On Deck	4 + 4
2,4-D + dichlorprop (2,4-DP)	Patron 170	4 + 4
2,4-D + clopyralid + dicamba	Millennium Ultra 2	4 + 4 + 4
2,4-D + dichlorprop (2,4-DP) + dicamba	Brushmaster, Super Trimec	4 + 4 + 4
2,4-D + fluroxypyr + dicamba	Escalade 2, E-2	4 + 4 + 4
2,4-D + fluroxypyr + halauxifen-methyl	GameOn	4 + 4 + 4
2,4-D + fluroxypyr + triclopyr + sulfentrazone	Momentum 4-Score	4 + 4 + 4 + 14
2,4-D + imazapyr + fluazifop + dicamba + pyraflufen-ethyl	Barrier Year-Long Vegetation Control Concentrate	4 + 2 + 1 + 4 + 14
2,4-D + MCPA + dicamba	Triad Select	4 + 4 + 4
2,4-D + MCPA + dicamba + sulfentrazone	Triad SFZ Select, Terradex Quick Strike	4 + 4 + 4 + 14
2,4-D + MCPA + triclopyr + dicamba	Triad T Select, Terradex Power Premix	4 + 4 + 4 + 4
2,4-D + mecoprop (MCPP) + dicamba	3-D, Eliminate LO, Eliminate-D, EndRun, MEC Amine-D, Strike 3, Threesome, Three-Way, Trimec 1000, Trimec 992, Trimec 899, Trimec Bentgrass Formula, Trimec Classic, Trimec LAF-637, Trimec Southern, Triplet Low Odor, Triplet SF, TruPower2, TruPower3, Vessel	4 + 4 + 4
2,4-D + mecoprop (MCPP) + dicamba + carfentrazone	SpeedZone EW, SpeedZone Southern EW	4 + 4 + 4 + 14
2,4-D + mecoprop (MCPP) + dicamba + pyraflufen ethyl	RedZone 2	4 + 4 + 4 + 14
2,4-D + mecoprop (MCPP) + dicamba + sulfentrazone	Surge, SureZone	4 + 4 + 4 + 14
2,4-D + mecoprop (MCPP) + dichlorprop (2,4-DP)	Spoiler, Triamine, Triple Threat	4 + 4 + 4
2,4-D + quinclorac + dicamba	2DQ, Gordon's Trimec Crabgrass Plus Lawn Weed Killer, Medusa, Momentum Q, Quincept, Terradex Crabgrass & Broadleaf, Triad QC Select	4 + 4 + 4
2,4-D + quinclorac + dicamba + sulfentrazone	Q4 Plus	4 + 4 + 4 + 14
2,4-D + triclopyr	Aquasweep, Chaser, Chaser 2 Amine, Crossroad, CrossCut Select, Everett, Turflon II Amine	4 + 4
2,4-D + triclopyr + dicamba + pyraflufen ethyl	4-Speed XT	4 + 4 + 4 + 14
2,4-D + triclopyr + dicamba + sulfentrazone	Foundation, Triad TZ Select, TZONE, TZONE SE	4 + 4 + 4 + 14
2,4-D + triclopyr + fluroxypyr	Momentum FX2	4 + 4 + 4
2,4-D + triclopyr + fluroxypyr + flumioxazin	Sure Power, Tetra	4 + 4 + 4 + 14
amicarbazone	Xonerate 2SC	5
asulam	Asulox, Addax	18
atrazine	AAtrex 4L, AAtrex Nine-O, Atrazine 4L, Atrazine 90DF	5
benefin	Balan 2.5G, Crabgrass Preventer	3
benefin + oryzalin	Surflan XL 2G, XL 2G	3 + 3
benefin + trifluralin	Crabgrass Control 2% Team, Team 2G, various fertilizers with Team Pro	3 + 3
bensulide	Bensumec 4LF, Pre-San Granular 12.5G, Weedgrass Preventer	0
bensulide + oxadiazon	Goosegrass/Crabgrass Control	0 + 14
bentazon	Basagran T/O Biscayne	6
bispyribac-sodium	Velocity PM	2
bromoxynil	Broclean, Brox 2EC, Buctril, Buctril 4EC, MOXY 2E	6
carfentrazone	Aim EC, Antik EC, Quicksilver T&O	14
carfentrazone + sulfentrazone	Dismiss NXT, Spartan Charge	14 + 14
chlorsulfuron	Chlorsulfuron 75, Telar XP	2
clopyralid	Lontrel	4

Common and Trade Names and Mechanism of Action of Registered Herbicides (*continued*)

Common Name	Trade Name(s)	Mechanism of Action[1]
dazomet	Basamid G	17
dicamba	Banvel, Clarity, Clash, Cruise Control, Detonate, Diablo, Doleac DMA, Dolerity DGA, Rifle, Sterling Blue, Strut, Topeka, Vanquish, Vision	4
dicamba + halosulfuron	Yukon, Daikon	4 + 2
dicamba + metsulfuron	Fahrenheit	4 + 2
dimethenamid	Tower	15
dimethenamid + pendimethalin	FreeHand 1.75G	15 + 3
diquat	Diquash, Diquat, Diquat SPC 2 L, Reward, Solera Diquat, Tsunami DQ	22
dithiopyr	CGC 2L, CGC 40, Dimension 2EW, Dimension Ultra 40WP, Dithiopyr, Dithiopyr 40WSB, various fertilizers with dithiopyr	3
dithiopyr + isoxaben	Crew	3 + 21
ethofumesate	Poa Constrictor, Prograss, Prograss SC, Thrasher	15
fenoxaprop	Acclaim Extra	1
fenoxaprop + fluroxypyr + dicamba	Last Call	1 + 4 + 4
flazasulfuron	Katana	2
florasulam	Defendor SC	2
fluazifop	Fusilade II, Ornamec 170	1
flumioxazin	Flumioxazin 51 WDG, Semera 51WDG, Semera SC, StayGuard, SureGuard WDG, Sureguard SC	14
fluroxypyr	Vista XRT, Flagstaff	4
foramsulfuron	Revolver	2
glufosinate	Finale, Finale XL T&O, Cheetah Pro, Surmise Pro, X-Out	10
glufosinate + pelargonic acid + imazethapyr	Onmark, Surmise SpeedPro XT	9 + 17 + 2
glyphosate	Aqua Neat, Aqua Star, Departure, Glyfos, Glyfos X-TRA, Glyphosate 4 Plus, Glyphosate 5.4, Glyphosate Plus, Glyphomate 41, Glypro Plus, GlyStar Original, GlyStar Plus, KleenUp Pro, Pronto, Prosecutor, Prosecutor Pro, Razor, Razor Pro, Refuge, Rodeo, Roundup Custom, Roundup Pro, Roundup Pro Concentrate, Roundup PROMAX, Roundup WeatherMax, Showdown, Touchdown Total	9
glyphosate + 2,4-D	Campaign	9 + 4
glyphosate + diquat	QuickPro, Razor Burn	9 + 22
glyphosate + prodiamine	ProDeuce	3 + 9
halosulfuron	Empero, Empero Q-Pak, Halo 5WDG Select, Halo 75WDG Select, ProSedge, Profine 75, Sandea, SedgeHammer, SedgeHammer+, SedgeMaster, Stadia	2
imazapic	Panoramic 2SL, Plateau	2
imazaquin	Image 70DG, Scepter	2
imazosulfuron	Celero	2
indaziflam	Specticle FLO, Specticle G, Specticle 20WSP	29
isoxaben	Gallery 75DF, Gallery SC, Isoxaben 75WG	21
MCPA	MCPA-4 Amine, MCPA Ester 4, Wildcard	4
MCPA + fluroxypyr + dicamba	Change Up	4 + 4 + 4
MCPA + fluroxypyr + dichlorprop	Chaser Ultra 2	4 + 4 + 4
MCPA + fluroxypyr + triclopyr	Battleship III	4 + 4 + 4
MCPA + mecoprop (MCPP) + dicamba	Trimec Encore, Tri-Power	4 + 4 + 4
MCPA + mecoprop (MCPP) + dicamba + carfentrazone	PowerZone	4 + 14
MCPA + triclopyr + dicamba	Cool Power, Eliminate, Horsepower, Spurge Power, Three-Way Ester II	4 + 4 + 4
mecoprop (MCPP)	MCPP-p 4 Amine, Mecomec 2.5, Mecomec 4	4
mesotrione	Meso 4SC Select, SlipStream, Tenacity, Torocity, TRIONE XL	27

Common and Trade Names and Mechanism of Action of Registered Herbicides (*continued*)

Common Name	Trade Name(s)	Mechanism of Action[1]
methiozolin	PoaCure	30
metolachlor	Pennant MAGNUM, Spirus	15
metribuzin	Sencor 75%, Metricor DF	5
metsulfuron	Manor, Mansion, MSM 250D, MSM Turf, Patriot, Rometsol	2
metsulfuron + rimsulfuron	Negate 37WG	2 + 2
MSMA	MSMA 6.6, MSMA 6 Plus, TARGET 6.6, TARGET 6 Plus, 912 Herbicide	17
oryzalin	Harrier 4L, Oryzalin 4, Surflan AS, Surflan WDG	3
oxadiazon	Oxadiazon 50WSB, Oxadiazon 2G, Oxadiazon SC, Oxa-Pro 2G, Ronstar 50WSP, Ronstar Flo, Starfighter L, various fertilizers with oxadiazon	14
pelargonic acid	Beloukha, Scythe	17
pendimethalin	Hammerkop, Pendulum 3.3EC, Pendulum 2G, Pendulum Aqua Cap, Pin-Dee 3.3T&O, Pre-M 3.3 EC, Pre-M Aquacap, Up-End, various fertilizers with pendimethalin	3
penoxsulam	LockUp, Sapphire	2
penoxsulam + 2,4-D + dicamba	LockUp Extra 2 + fertilizer	2 + 4
penoxsulam + 2,4-D + dicamba + dithiopyr	LockUp plus Dimension + fertilizer	2 + 4 + 4 + 3
penoxsulam + sulfentrazone + 2,4-D + dicamba	Avenue South	2 + 14 + 4 + 4
pinoxaden	Manuscript	1
prodiamine	Barricade 4FL, Barricade 65WG, Basilisk, Cavalcade 65WDG, eVade 4FL, Kade 65WDG, Knighthawk, LESCO Stonewall 65WDG, ProClipse 65WDG, Prodiamine 4L, Prodiamine 4SC Select, Prodiamine 65WG, Resolute 4FL, Resolute 65WG, various fertilizers with prodiamine	3
prodiamine + isoxaben	Gemini G, Gemini 3.7 SC	3 + 21
prodiamine + quinclorac	Cavalcade PQ, LESCO Stonewall PQ	3 + 4
pronamide	Kerb 50WP, Kerb SC T&O	3
pyraflufen ethyl	Octane 2% SC	14
pyrimisulfan	Arkon, Vexis	2
quinclorac	Acclaim Accelerate, Drive XLR8, Eject 4L, Eject 75DF, Q-Ball, Quinclorac 1.5L, Quinclorac 1.5L Select, Quinclorac 75 DF Select, Quinclorac SPC 75DF, QuinPro Herbicide, Quintessential, Quin-Way	4
quinclorac + sulfentrazone	Solitaire, Solitaire WSL	4 + 14
rimsulfuron	Rimsulfuron 25DF	2
sethoxydim	Segment, Segment II, Sethoxydim SPC	1
simazine	Princep, Sim-Trol 4L, Sim-Trol 90DF, Simazine 4L, Simazine 90DF	5
simazine + prodiamine + imazaquin	Coastal	5 + 3 + 2
sulfentrazone	Dismiss, Acquit, Aquestra, Expel, Spartan 4F, SULF 396, Sulfentrazone 4L Select, Sulfentrazone 4SC Select, SUREPYC	14
sulfentrazone + imazethapyr	Dismiss South	14 + 2
sulfentrazone + metsulfuron	Blindside	14 + 2
sulfentrazone + prodiamine	Echelon 4SC, various fertilizers with sulfentrazone + prodiamine	14 + 3
sulfosulfuron	Certainty, Sertay	2
thiencarbazone + foramsulfuron + halosulfuron	Tribute TOTAL	2 + 2 + 2
thiencarbazone + iodosulfuron + dicamba	Celsius WG	2 + 2 + 4
thiencarbazone + iodosulfuron + halosulfuron	Celsius XTRA	2 + 2 + 2
topramezone	Pylex	27
triclopyr	Boulder 6.3, SightLine, Turflon Ester Ultra, Triclopyr 4, Trycera	4
triclopyr + clopyralid	Confront, 2-D	4 + 4
triclopyr + dicamba + mesotrione	Sublime	4 + 4 + 27
triclopyr + fluroxypyr	Tailspin	4 + 4
trifloxysulfuron	Monument 75WG, Recognition	2

[1] For each herbicide, the number is a descriptive code for the mechanism of action that follows. See *Herbicide Mechanism of Action* on page 110 for more information.

Herbicide Mechanism of Action

The Weed Science Society of America (WSSA) developed a classification system to identify the biochemical site where an herbicide works to kill a weed. The WSSA system uses a numerical code, which is placed at the top of the herbicide label. For example, 4 is the group number for a synthetic auxin herbicide.

GROUP 4 HERBICIDE

Herbicide Movement

After a foliar application, herbicides move (translocate) within a plant in one of three ways:

1. Symplastically translocated (herbicides move to growing points or storage organs)

2. Apoplastically translocated (herbicides only capable of upward movement)

3. Contacts (herbicides do not move appreciably but kill very quickly)

Herbicides that move to growing points or storage organs are symplastically translocated. The symplast is the portion of the plant that carries water and solutes such as sugars. Herbicides in this class often take several days to produce symptoms or kill the weed. Examples include: 2,4-D, glyphosate (Roundup), halosulfuron (SedgeHammer), and triclopyr (Turflon Ester Ultra). These herbicides are the best choice when the goal is to kill underground portions (that is, rhizomes, taproots, etc.) of perennial weeds.

Herbicides that only move upward in plant tissues are apoplastically translocated. The apoplast is the portion of the plant that carries water and minerals. Plant roots or lower leaves often take up these herbicides, which then move to the leaf tips. When applied to the foliage, these herbicides have limited movement since water flows up from the roots and out to the leaves. Examples include amicarbazone (Xonerate 2SC), bentazon (Basagran), and simazine (Princep).

Some herbicides kill weeds with limited absorption or limited movement in the plant. Other herbicides are absorbed but kill plant cells rapidly, which limits their movement in plant tissues. These herbicides that have limited redistribution in plant tissues are called contact or nontranslocated herbicides. Examples include carfentrazone (Quicksilver), diquat (Reward), glufosinate (Finale), and sulfentrazone (Dismiss). Since these herbicides do not move throughout the plant tissue, excellent spray coverage is necessary for weed control.

Not all herbicides are applied to the foliage. Some herbicides are soil-applied and work as shoot and/or root inhibitors. These herbicides have limited or no foliar activity and are primarily used for preemergence control of annual grassy and broadleaf weeds. They may be applied to foliage, but they are not effective until they move into the soil after rainfall or irrigation. Examples include indaziflam (Specticle), pendimethalin (Pendulum), and prodiamine (Barricade).

Herbicide Classification According to Movement and Mechanism of Action

This table classifies herbicides according to their WSSA group number.

Mechanism of Action	Herbicide Movement	WSSA Group Number
specific site unknown	unknown	0
acetyl CoA carboxylase (ACCase) inhibitors	symplast	1
acetolactate synthase (ALS) inhibitor	symplast	2
microtubule inhibitors	soil-applied	3
synthetic auxins	symplast	4
photosystem II inhibitors	apoplast	5, 6
enolpyruvyl shikimate-3-phosphate (EPSP) synthase inhibitor	symplast	9
glutamine synthetase inhibitor	contact	10
protoporphyrinogen oxidase (PPO or Protox) inhibitors	contact	14
very long chain fatty acid inhibitor	soil-applied	15
potential nucleic acid inhibitors or nondescript mode of action	symplast	17
dihydropteroate synthase inhibitor	symplast	18
photosystem I inhibitors	contact	22
4-hydroxyphenylpyruvate dioxygenase (HPPD) inhibitor	symplast	27
cellulose biosynthesis inhibitors	soil-applied	21, 29
inhibition of fatty acid thioesterase	apoplast	30

Plant Growth Regulators

Plant Growth Regulators Used for Turf: Their Action and Uptake

Common Name	Type	Class[1]	Site of Action	Uptake	Seedhead Suppression	Trade Names
trinexapac-ethyl	II	A	Inhibits GA late in biosynthesis	Foliar	No	GameUp, Governor G, Groom PGR, Podium, PGR 113 MC, Pramaxis MEC, Primo Maxx, RegiMax PGR, T-Pac SPC MEC, Trin-Pac Select, Trinexapac-Ethyl 1AQ
prohexadione-Ca	II	A	Inhibits GA late in biosynthesis	Foliar	No	Acuity, Anuew, Anuew EZ
paclobutrazol	II	B	Inhibits GA early in biosynthesis	Root	No	Griffin, Travek 2SC, Trimmit 2SC, PAC 223, Tide Paclo 2SC, Turf Enhancer 2SC
flurprimidol NY	II	B	Inhibits GA early in biosynthesis	Root	No	Cutless 50W, Cutless MEC
flurprimidol + trinexapac-ethyl NY	II	B + A	Inhibits GA biosynthesis	Root and foliar	No	Legacy, Edgeless
flurprimidol + paclobutrazol + trinexapac-ethyl NY	II	B + A	Inhibits GA biosynthesis	Root and foliar	No	Musketeer
maleic hydrazide L	I	C	Not well understood	Foliar	Yes	Royal MH-30 SG, Retard
ethephon		E	Increases ethylene synthesis	Foliar	Yes	Proxy, Cardinal, Ethephon 2, Ethephon 2SL, Oskie, Verve
gibberellic acid[2] W		E	Increases gibberellic acid content	Root and foliar	No	ProGibb T&O, RyzUp SmartGrass

1 Class A plant growth regulators (PGRs) interfere with the production of gibberellins late in the biosynthetic pathway. Class B PGRs interfere with the production of gibberellins early in the biosynthetic pathway. Class C PGRs prevent cell division (mitotic inhibitors) for a defined period and provide excellent seedhead suppression. Class D products are PGRs at very low rates but function as herbicides at higher rates. Class E PGRs are classified as plant hormones. Class D and F PGRs not listed.

2 The only PGR that accelerates leaf elongation.

W This product is labeled for use primarily in warm-season turf. Read the label for a list of tolerant grasses.

NY Use or sale of this product has restrictions in New York State or Long Island, New York (Nassau and Suffolk counties). Refer to the label for more information.

L This product is registered only in specific North Central states. Check with local contacts for availability.

Plant Growth Regulators for General Turf Use

Refer to product labels for specific turfgrass species use instructions.

Trade Name (product/A)	Common Name (lb ai/A)	Turfgrasses	Comments
Anuew (1.8–24.0 oz/A) Anuew EZ (2.2 to 52.4 fl oz/A)	prohexadione calcium (0.017–0.41)	bermudagrass, creeping bentgrass, Kentucky bluegrass, perennial ryegrass	Use on actively growing turf to suppress growth and manage clippings. Rate varies by turf species. Foliar-absorbed and rainfast within 1 hour. Anuew provides a similar magnitude of growth suppression as trinexapac-ethyl but is observed to last about 50% longer on bentgrass putting greens. Clipping reduction should be evident within 3 days of an application. Use a lower application rate or delay application if stressful growing conditions are imminent. Wait at least 1 hour after mowing to apply. Do not mow the treated area for 24 hours after application. The liquid formulation (Anuew EZ) is formulated at 1 pound active ingredient per gallon. A 12 oz/A rate of Anuew (WDG) applies an equivalent amount of ingredient to a 26.4 fl oz/A rate of Anuew EZ.
Cutless MEC (6.1–73.8 fl oz) Cutless 50W (2–24 oz) NY	flurprimidol (0.06–0.75)	bermudagrass, creeping bentgrass, Kentucky bluegrass, perennial ryegrass, tall fescue, zoysiagrass	Apply to actively growing cool-season turf species to suppress annual bluegrass, or to suppress growth and manage clippings. Do not apply to areas where annual bluegrass (*Poa annua*) is the desired turfgrass. Apply in the spring after 3–4 mowings or in the fall. If necessary, repeat the application at 3-6-week intervals. Do not exceed 3.0 lb ai/A per growing season. Make the final fall application 8 weeks before the onset of winter dormancy. Delay reseeding for 2 weeks after application. Do not use on sod farms. See label for additional instructions. On bermudagrass and zoysiagrass: Cutless may be used on medium- to high-quality, well-maintained turfgrasses. Refer to label for rates for different bermudagrass cultivars and zoysiagrass. Make the first application after full spring green-up (usually mid-May to mid-June). An additional application may be made in late summer to bermudagrass only. Treated areas should receive 0.5 inch of rainfall or irrigation within 24 hours of application. Make uniform application with a boom-type sprayer and avoid overlaps.
Legacy (5–30 fl oz) Edgeless (30–60 fl oz) NY	flurprimidol + trinexapac-ethyl	bermudagrass, creeping bentgrass, Kentucky bluegrass, perennial ryegrass, zoysiagrass	Use on actively growing turf to suppress annual bluegrass, or use to suppress growth and manage clippings. Apply the 5 fl oz/A rate if more than 50% annual bluegrass is present in creeping bentgrass putting greens. Apply the 10 fl oz/A rate if < 50% annual bluegrass is present in creeping bentgrass putting greens. Higher rates (15–30 fl oz/A) can be used on Kentucky bluegrass and perennial ryegrass. Rates of 8–15 fl oz/A should be used on bermudagrass and zoysiagrass. Spray interval will vary 2–6 weeks depending on desirable growth suppression, rate, and other factors. Do not seed 3 weeks before or after application. Do not apply to turf under temperature, moisture, or pest stress. Do not exceed 350 fl oz/A per year of Legacy. Apply in the spring after turf resumes active growth. Make last application 4 weeks before the onset of inactive grass growth. Only use Edgeless formulation for edging applications.

NY Use or sale of this product has restrictions in New York State or Long Island, New York (Nassau and Suffolk counties). Refer to the label for more information.

Plant Growth Regulators for General Turf Use (*continued*)

Trade Name (product/A)	Common Name (lb ai/A)	Turfgrasses	Comments
Musketeer (12–30 fl oz) NY	flurprimidol + paclobutrazol+ trinexapac-ethyl (see label)	bermudagrass, creeping bentgrass, Kentucky bluegrass, perennial ryegrass	Use on actively growing, well-established turf to suppress annual bluegrass, or use to suppress growth and manage clippings. Apply the 12–15 fl oz/A rate if more than 50% annual bluegrass is present in creeping bentgrass putting greens. Apply the 12–18 fl oz/A rate if < 50% annual bluegrass is present in creeping bentgrass putting greens. Higher rates (18–30 fl oz/A) can be used on Kentucky bluegrass and perennial ryegrass. Rates of 15–30 fl oz/A should be used on bermudagrass. Spray interval will vary 2–6 weeks depending on desirable growth suppression and rate, as well as other factors. Do not seed 2 weeks before or after application. Do not apply to turf under temperature, moisture, or pest stress. Do not exceed 580 fl oz/A per year of Musketeer. Apply in the spring after turf resumes active growth. Make last application 4 weeks before the onset of inactive grass growth or winter dormancy. Do not use on residential turf. Do not use within 4 weeks of a preemergence herbicide application.
Primo MAXX, T-NEX 1AQ, and many other 1 MEC formulations (see label) Governor G (see label)	trinexapac-ethyl (see label)	annual bluegrass, bermudagrass, buffalograss, creeping bentgrass, fine fescue, Kentucky bluegrass, perennial ryegrass, tall fescue, zoysiagrass	Use on actively growing turf to suppress growth and manage clippings. Rate varies by turf species. Foliar-absorbed and rainfast within 1 hour. Make applications to cool-season grasses in the spring and early summer at 4-week intervals to help with clipping management. A typical program for a bermudagrass fairway would start in June and repeat at 4-week intervals through the peak growing season. Do not mow < 1 hour before or < 4 hours after application to minimize injury. To extend the life of painted lines on sports fields, add 1 fl oz of Primo MAXX, or 0.5 oz Primo 25 WSB to each gallon of marking paint. This will extend the life of lines 7–14 days on cool-season and up to 30 days on warm-season grasses. A lower rate of trinexapac-ethyl may be used in athletic field paint depending on the paint application rate.
Proxy, Ethephon 2SL, and many other 2 SL formulations (1.7 gal/A)	ethephon (see label)	annual bluegrass, creeping bentgrass, fine fescue, Kentucky bluegrass, perennial ryegrass, tall fescue	Not for residential use. Useful for annual bluegrass seedhead suppression. Make the initial application before new seedheads emerge. Do not exceed 30 fl oz/1,000 ft² per year.
ProGibb T&O (1.0–25.0 fl oz) RyzUp Smartgrass (0.3 to 1.0 oz) W	gibberellic acid (see label)	bermudagrass	**Do not use ProGibb T&O on cool-season turf.** Apply to promote growth and reverse the effects of gibberellic acid-inhibiting plant growth regulators such as trinexapac-ethyl. This PGR is also used to prevent discoloration during periods of cold stress and light frost on bermudagrass.
Royal MH-30 SG (1.25 to 10.0 lb/A) Retard (0.25 to 1.3 gal/A) L	maleic hydrazide (see label)	bermudagrass, creeping bentgrass, fine fescue, Kentucky bluegrass, perennial ryegrass, tall fescue, zoysiagrass	This product reduces leaf growth and seedhead formation. Use in the spring or fall, but do not apply in summer. Not recommended for areas where aesthetic appearance is more important than reducing mowing requirements.
Trimmit 2SC, Tide Paclo 2SC, and many other 2SC formulations (see label) Turf Enhancer (see label on fertilizer carrier or 2SC formulation) TGR (see label, on fertilizer carrier)	paclobutrazol (see label)	bermudagrass, creeping bentgrass, fine fescue, Kentucky bluegrass, perennial ryegrass, tall fescue, zoysiagrass	Not for residential use. Use on actively growing turf to suppress annual bluegrass or to suppress growth and manage clippings. Spring and fall timings are recommended for annual bluegrass (*Poa annua*) suppression. Apply when annual bluegrass is actively growing. Repeated applications can be made in the spring until approximately 1 month before the onset of high air temperatures. In the fall, apply at least 1 month before the first frost. Make applications in 2–3 gal spray volume per 1,000 ft² for best results. Water-in with at least 0.25 inch of irrigation within 24 hours of application. Do not use on areas with > 70% *Poa annua*. This program is designed to encourage a gradual transition from *Poa annua* to bentgrass and will take repeated applications over several years. Do not exceed 2.0 lb ai/A per year.

W This product is labeled for use primarily in warm-season turf. Read the label for a list of tolerant grasses.
NY Use or sale of this product has restrictions in New York State or Long Island, New York (Nassau and Suffolk counties). Refer to the label for more information.
L This product is registered only in specific North Central states. Check with local contacts for availability.

Successfully Using Plant Growth Regulators in Turf

Plant growth regulators (PGRs) include any product that alters plant growth and development—the products can be naturally derived or synthetic. PGRs include plant hormones, plant or microbial extracts, and herbicides. Fertilizers are not considered PGRs despite their effect on growth rate.

In turf, the most commonly applied PGRs inhibit the production of the phytohormone gibberellin (GA). This hormone increases the rate of leaf elongation in mature turfgrass stands. By slowing GA production, PGRs can reduce clipping yield production and mowing requirements, which is the primary benefit of PGRs. These PGRs are classified as Class A (prohexadione-Ca and trinexapac-ethyl) or Class B (flurprimidol and paclobutrazol), depending on the product's site of action and uptake. This will be the focus of this section.

Many PGR products can have favorable secondary benefits. They can increase turfgrass quality, improve stand density, enhance shade and drought stress tolerance, reduce nutrient requirements, suppress seedheads, and suppress weeds like annual bluegrass. A majority of golf course superintendents use PGRs to maintain putting green speed; however, research shows that this response can be inconsistent or minimial on cool-season greens. The secondary benefits only occur when superintendents sustain growth suppression throughout the growing season.

General Growth Response to PGRs

The GA-inhibiting PGRs typically affect clipping yield in two phases. Clipping yield will decline relative to untreated turf following the PGR application. Suppression will intensify for a period before it wanes, and clipping yield is enhanced relative to untreated turf. This enhanced growth rate (called the "rebound phase") should be avoided because many of the secondary benefits associated with PGR use cease (such as improved stress tolerance and green color). The amount (intensity) and duration of clipping yield suppression depends on the active ingredient and rate of the PGR, the turfgrass species, and management. The duration of suppression is also closely related to air temperature. When air temperatures are high, growth suppression periods are shorter than when air temperatures are cooler. As a result, growing degree day (GDD) accumulation following the most recent PGR application can accurately predict the growth response and when a PGR needs to be reapplied (See *Weather Effects on Applications,* page 59 and *Plant Growth Regulator GDD Calculation Math,* page 121).

Here are some important points to consider about PGR growth response and GDD models for PGRs:

- For cool-season turfgrass species, higher mowed turf typically has longer and more intense clipping yield suppression than putting green turfgrass.
- The application rate generally has a much larger effect on the amount of growth suppression but not the duration of suppression. For example, paclobutrazol applied to a creeping bentgrass green at 5.5 fl oz/A typically suppresses yield by 31%, while the 16 fl oz/A rate typically suppresses yield by 53%. Interestingly, the ideal reapplication interval for the 5.5 and 16 fl oz/A application rate is 270 and 310 GDD (base 0°C), respectively. That's a difference of less than two calendar days during most summers.
- Growing degree day (GDD) models use temperature accumulation to predict the duration of growth suppression on specific turfgrass species under different growing environments. Experiments have established thresholds to help guide reapplication scheduling. These thresholds are also specific to management (putting greens vs. fairways), product, and product rate (in some cases). The GDD models most effectively predict the duration of suppression. The amount of suppression is more variable (especially once growth suppression wanes) and depends on various external factors (such as chronic heat or moisture stress) and even turfgrass cultivar.
- GDD models for PGRs use temperatures measured in Celsius with a base temperature of either 0°C for cool-season turf or 10°C for warm-season species. The models start the first day a PGR is applied to an area and are reset to 0 GDD the morning the PGR is applied. Using Fahrenheit temperatures will result in applying PGRs too frequently.
- A web-based application to help manage PGR applications is available at *https://greenkeeperapp.com/.* It automatically suggests the ideal GDD interval based on site, species, PGR, and application rate. GreenKeeper also tracks GDD accumulation, with the appropriate base temperature, from data generated by the nearest weather station.
- Reapplying PGRs at half the ideal GDD interval can intensify clipping yield suppression. For example, applying trinexapac-ethyl to a creeping bentgrass green at 5.5 fl oz/A every 200 GDDs (base 0°C) reduced yield by 18%. That same application rate applied every 100 GDDs increased suppression to 35%.

- The rebound phase does not always occur. Observations suggest the potential for a large rebound is greatest in the spring and early summer. The rebound phase is generally smaller in late summer and during intense heat. A rebound phase is most frequent on cool-season golf course putting greens compared to higher mowed turf at fairway or lawn height cut.
- Nitrogen mineralization frequently causes accelerated growth in mid- to late summer. Many turfgrass managers blame these mid-season growth surges on PGR rebound. In reality, the growth rate would be even faster without a PGR during these times. Use GDD intervals to sustain season long growth suppression.

PGR Application to Golf Course Putting Greens Complexes

PGRs are most frequently applied to golf course putting greens. Reapplication intervals are shorter on putting greens than fairways because putting greens are mowed daily and have a faster growth rate. This results in significant PGR removal during mowing. The surrounding collar, approach, and rough have much longer growth suppression and are more sensitive than the putting green turf. Frequent PGR overspray on collars and surrounds can lead to excessive levels of growth suppression and collar decline.

Here are some facts about applying PGRs to cool-season golf course putting greens and green surrounds:

- The application rate response for Class A PGRs (prohexadione-Ca and trinexapac-ethyl) is minimal. For prohexadione-Ca, the ideal reapplication interval is 250 GDDs (base 0°C), while the ideal reapplication interval for trinexapac-ethyl is 200 GDDs (base 0°C). Both PGRs suppress clipping yield by approximately 20% to 30% depending on species and cultivar.
- Products that contain Class B PGRs (flurprimidol and paclobutrazol) are more responsive to application rate than Class A PGRs. Higher application rates have a much greater effect on the amount or intensity of the growth suppression phase than the duration of suppression. Expect 20% to 50% growth suppression with application intervals ranging from 210 to 310 GDDs, depending on active ingredient and application rate.
- Roots absorb Class B PGRs so should ideally be watered in lightly following application. This can be problematic because putting green sprays can contain other products that should not be watered in. Also, some PGRs contain both Class A and Class B active ingredients (Legacy and Musketeer). For these situations, irrigating the night following application usually results in comparable growth suppression to watering in the PGR immediately after application.
- Applying PGRs on putting green intervals to collars and surrounds can intensify growth suppression (that is, overregulation). This can cause discoloration, grass species/cultivar segregration, and limited traffic recovery. For example, applying trinexapac-ethyl or paclobutrazol to bentgrass mowed at 0.400 inch resulted in 70% to 95% growth suppression after four successive applications (trinexapac-ethyl applied every 200 GDDs or paclobutrazol applied every 260 GDDs).
- Using GPS sprayer technology can reduce overspray of PGRs on collars and surrounds. However, PGRs are rarely applied alone and this strategy may limit control products and fertilizers from being applied to collars. Another approach is to not apply PGRs to putting greens because it is possible to produce high-quality turf without using PGRs.
- Some managers have reduced the mowing height and increased mowing frequency on their collars. This accelerates PGR removal from mowing and lessens the risk of developing PGR overregulation.

Plant Growth Regulator Suppression and Suggested Reapplication Intervals

Refer to product labels for specific turfgrass species use instructions.

Trade Name (product/A)	Common Name	Creeping Bentgrass Putting Green 0.120" Height of Cut			Creeping Bentgrass Collar or Fairway 0.400" Height of Cut		
		Rate Range Evaluated	Peak Suppression (%)	Reapplication Interval (in GDDs)	Rate Range Evaluated	Peak Suppression (%)	Reapplication Interval (in GDDs)
Anuew EZ	prohexadione calcium	4.5-17.6 fl oz/A	20-25	250-280[1]	9-27 fl oz/A	60-75	350-380
Cutless MEC (Cutless 50W)	flurprimidol	6.1-24.6 fl oz/A (2.0-8.0 oz/A)	20-30	180-240	25-49 fl oz/A (8-16 oz/A)	30-40	380-410
Legacy	flurprimidol + trinexapac-ethyl	5-10 fl oz/A	20-35	240-270	10-20 fl oz/A	40-70	320-390
Musketeer	flurprimidol + paclobutrazol+ trinexapac-ethyl	12-22 fl oz/A	25-40	255-265	18-30 fl oz/A	55-70	350-400
Primo Maxx	trinexapac-ethyl	5.5-11.0 fl oz/A	20-25	200-230	11-33 fl oz/A	55-80	350-380
Trimmit 2SC	paclobutrazol	5.4-16.3 fl oz/A	30-50	250-280	16-32 fl oz/A	70-80	480-640

1 Growing degree day (GDD) intervals represent the recommended GDD interval to sustain season long yield suppression. GDDs are calculated in degrees Celsius starting the day of the most recent PGR application. The base temperature is 0°C for all cool-season turfgrass species. The models are reset the morning the next PGR is applied.

PGR Overapplication and Overregulation

Many mixed turf sites (like golf course putting green complexes) are prone to excessive PGR growth suppression. This can result from PGR applications that are too frequent or rates that are too high. A severely limited growth rate can reduce traffic tolerance, promote disease, and slow recovery from pests. High PGR application rates can also increase the risk of turf phytotoxicity.

Here are some tips to minimize the risk or damage from over application of too much PGRs:

- Since trinexapac-ethyl (Primo Maxx) was introduced in 1993, PGR use in managed turf has significantly increased. Applying GA-inhibiting PGRs brings both primary (reduced clipping yield and mowing) and secondary benefits (increased turf quality, density, etc.), but turf managers must remember that their primary function is to reduce growth. A common mistake is to apply PGRs at too high a rate on athletic fields and other turf areas, which results in reduced recovery rates from injuries. For this reason, use PGRs cautiously and at low rates in trafficked turf.
- A single, mistakenly high application rate of GA-inhibiting PGR is rarely lethal. In those cases, masking phytototoxicity with iron or a turf colorant can be effective. Wait until the growth has returned to a normal level before restarting the PGR program.
- Higher application rates of nitrogen fertilizer can help with overregulation from too frequent PGR applications (collars and approaches). Adding 0.2–0.4 lb of soluble nitrogen per 1,000 ft^2 can quickly stimulate growth and recovery. However, the risk of rebound is also greater if the PGR program would be stopped in the middle of the season.
- Applying products that contain the phytohormone GA can rescue turf from overregulation. However, be careful with such products because very low rates can cause rapid etiolation in turf that does not have overregulation. Etiolation is a rapid elongation of the leaves that occurs overnight and often results in yellow or chlorotic turf. Applying a GA with RyzUp SmartGrass at 0.1 to 0.2 oz/A or ProGibb T&O at 1.0 to 2.0 fl oz/A will stimulate recovery. This RyzUp SmartGrass rate is below the manufacturer's suggested rate.

Annual Bluegrass Suppression in Creeping Bentgrass Putting Greens with Plant Growth Regulators (PGRs)

Refer to product labels for weeds controlled and specific turfgrass species use instructions.

Trade Name (product/A)	Common Name (lb ai/A)	Primary Purpose	Comments
Cutless MEC (6.1–24.6 fl oz/A) Cutless 50 W (2–8 oz/A)	flurprimidol (0.0625–0.25)	annual bluegrass suppression, growth supression	Make initial application after bentgrass putting greens are growing vigorously and have been mown 3–4 times. Repeat application at 2–4 week intervals. This program is designed to encourage a gradual transition from *Poa annua* to bentgrass and will take repeated applications over several years. Do not apply to putting greens other than those where bentgrass is the desired species. Do not use on areas with > 80% *Poa annua*.
NY Legacy (5–10 fl oz/A)	flurprimidol + trinexapac-ethyl (0.04 + 0.02–0.09 + 0.04)	annual bluegrass suppression	Apply in early spring and late fall at 5–10 oz/A and repeat application at 2–4 week intervals. Use lower rates on putting greens comprised of more than 50% annual bluegrass (*Poa annua*). This program is designed to encourage a gradual transition from *Poa annua* to bentgrass and will take repeated applications over several years. Do not overseed within 3 weeks following final application.
NY Musketeer (12–30 fl oz)	flurprimidol + paclobutrazol+ trinexapac-ethyl (see label)	annual bluegrass suppression, growth suppression	Use on actively growing, well-established turf to suppress annual bluegrass, or use to suppress growth and manage clippings. Apply the 12-15 fl oz/A rate if more than 50% annual bluegrass is present in creeping bentgrass putting greens. Apply the 12-18 fl oz/A rate if < 50% annual bluegrass is present in creeping bentgrass putting greens. Do not seed 2 weeks before or after application. Do not apply to turf under temperature, moisture, or pest stress. Do not exceed 580 fl oz/A per year of Musketeer. Apply in the spring after turf resumes active growth. Make last application 4 weeks before the onset of inactive grass growth or winter dormancy. Do not use on residential turf. Do not use within 4 weeks of a preemergence herbicide application.
NY Trimmit 2SC, Tide Paclo 2SC, and many other 2 SC formulations (6.4–16.0 fl oz) Turf Enhancer (see label on fertilizer carrier or 2SC formulation) TGR (see label, on fertilizer carrier)	paclobutrazol (0.1–0.25)	annual bluegrass suppression	Apply in spring and fall to suppress annual bluegrass (*Poa annua*). Apply in the spring after green-up and after the turf has been mown twice. Repeat applications in the spring until approximately 1 month before the onset of high air temperatures. In the fall, apply at least 1 month before the first frost. Make applications in 2–3 gal/1,000 ft^2 for best results. Water-in with at least 0.25 inch of irrigation within 24 hours of application. Do not use on areas with > 70% *Poa annua*. This program is designed to encourage a gradual transition from *Poa annua* to bentgrass and will take repeated applications over several years. Do not exceed 2.0 lb ai/A per year.

NY Use or sale of this product has restrictions in New York State or Long Island, New York (Nassau and Suffolk counties). Refer to the label for more information.

Pesticide and Plant Growth Regulator Math

Common Weights and Measures

Area

1,000 square feet = 92.9 square meters

1 acre	=	43,560 square feet	=	4,046.9 square meters	=	0.4047 hectare

Length

1 foot	=	12 inches	=	30.48 centimeters	=	0.3048 meters
1 yard	=	3 feet	=	91.44 centimeters	=	0.9144 meter
1 mile	=	1,760 yards	=	5,280 feet	=	1.61 kilometers

Weights

1 gram	=	1,000 milligrams		
1 ounce	=	28.35 grams		
1 pound	=	16 ounces	=	454 grams
1 kilogram	=	1,000 grams	=	2.205 pounds
1 ton	=	2,000 pounds	=	907 kilograms

Liquid Measures

1 teaspoon	=	0.333 tablespoons	=	0.1667 fluid ounce	=	4.93 milliliters						
1 tablespoon	=	3 teaspoons	=	0.5 fluid ounce	=	14.8 milliliters						
1 fluid ounce	=	2 tablespoons	=	6 teaspoons	=	29.6 milliliters						
1 cup	=	8 fluid ounces	=	16 tablespoons	=	236.6 milliliters						
1 pint	=	16 fluid ounces	=	2 cups	=	473.2 milliliters						
1 quart	=	32 fluid ounces	=	2 pints	=	4 cups	=	0.946 liter				
1 liter	=	2.113 pints	=	1.057 quarts	=	1,000 milliliters						
1 gallon	=	128 fluid ounces	=	4 quarts	=	8 pints	=	16 cups	=	3,785 milliliters	=	3.785 liters

Ounces or Ounces

When calculating the amount of pesticide needed to treat an area, the first question you need to ask is, "Do I have a dry or liquid product?" Products are available in many formulations. However, each formulation can be categorized as either a dry or liquid.

Dry *(measured as ounces, 16 oz = 1 pound)*	**Liquid** *(measured as fluid ounces, 128 fl oz = 1 gallon)*
DF = dry flowable	CS = capsulated suspension
G = granular	E = emulsifiable concentrate
SG = soluble dispersable granule, water-soluble granule	EC = emulsifiable concentrate
W = wettable powder	EW = emulsifiable in water
WDG = water dispersible granules	F = flowable
WG = water dispersible granules	FL = flowable
WSB = water-soluble bag	FLO = flowable
WSP = water soluble powder	L = liquid
	LF = liquid flowable
	ME = microemulsion
	OD = oil dispersion, oil-based suspension concentrate
	S = solutions
	SC = suspension concentrate
	SE = suspoemulsion
	WSL = water-soluble liquid
Scales are best used to determine the number of dry ounces (oz) of product needed. Measuring cups that come with dry fertilizers can sometimes be inaccurate.	Liquid measuring cups are best used to determine the number of fluid ounces (fl oz) of product needed.
When the product is a dry formulation, active ingredient is expressed as % active ingredient (ai).	When the product is a liquid formulation, active ingredient is expressed as (pounds) lb active ingredient (ai) per gallon (gal).
Example: Barricade 65WG	Example: Barricade 4FL
Barricade® 65WG	Barricade® 4FL
65% active ingredient (0.65 lb ai per 1 lb product), water dispersible granule	4 lb active ingredient per gallon, flowable

Amount of Product Needed

How much pesticide is needed to treat an area at a given rate?

Rate	Area to be treated	Conversion factors
1,000 ft^2		

or

Rate	Area to be treated	Conversion factors
A		

General formula: **Note:** the conversion factors here are the steps needed to convert to your final desired units (such as converting from gallons to fluid ounces, or from quarts to fluid ounces).

Example 1.

How many fluid ounces of Trimec 992 are needed to treat 0.4 acre of utility turf at a rate of 4 pints per acre?

4 pt	0.4 A	16 fl oz
1 A		1 pt

Multiply each value across the top row, then divide by each value on the bottom row. This should be computed as

4 pt × 0.4 A × 16 fl oz ÷ 1 A ÷ 1 pt =

Answer = 25.6 fl oz of Trimec 992

Amount to Add to the Spray Tank

How much product do we add to the tank?

General formula:

Tank size (gal)	A	Rate	Conversion factors
	Calibrated spray volume (gal)	A	

Example 2

How many ounces of Trimec 992 do we add to each spray tank? We have a 150-gallon spray tank that is calibrated to spray 50 gallons per acre. We need to treat 20 acres of turf with 4 pints of Trimec 992 per acre.

150 gal	1 A	4 pt	16 fl oz
	50 gal	1 A	1 pt

This should be computed as

150 gal × 1 A × 4 pt × 16 fl oz ÷ 50 gal ÷ 1 A ÷ 1 pt =

Answer = 192 fl oz (1.5 gal or 6 qt or 12 pt) of Trimec 992

How Many Tanks (trips with my sprayer) Does It Take?

How many tanks will it take me to treat this area?

General formula:

Calibrated spray volume (gal)	1 tank	Total area (A)
A	Tank size (gal)	

Example 3

How many full tanks will it take to treat 20 acres of turf with Trimec 992? We have a 150-gallon spray tank that is calibrated to spray 50 gallons per acre. We need to treat 20 acres of turf with 4 pints of Trimec 992 per acre.

50 gal	1 tank	20 A
1 A	150 gal	

This should be computed as

50 gal x 1 tank x 20 A ÷ 1 A ÷ 150 gal =

Answer = 6.7 tanks of Trimec 992

What If the Recommended Rate Is in Pounds of Active Ingredient?

How much pesticide is needed to treat an area at a given rate of active ingredient (ai)?

General formula (dry):

Rate	1 lb	Area to be treated	Conversion factors
A	% ai written as a decimal		

General formula (liquid):

Rate	1 gal	Area to be treated	Conversion factors
A	lb ai		

Example 4

How many ounces of Dimension Ultra 40WP are needed to treat 10,000 ft^2 of Kentucky bluegrass turf at 0.5 pounds of ai per acre?

0.5 lb ai	1 lb	10,000 ft^2	1 A	16 oz
A	0.40 lb ai		43,560 ft^2	1 lb

This should be computed as

0.5 lb ai × 1 lb × 10,000 ft^2 × 1 A × 16 oz ÷ 1 A ÷ 0.4 lb ai ÷ 43,560 ft^2 ÷ 1 lb =

Answer = 4.6 oz of Dimension Ultra 40WP

Example 5

How many fluid ounces of Dimension 2EW are needed to treat 2.0 acres of creeping bentgrass fairway at 0.375 pound of ai per acre?

0.375 lb ai	1 gal	2.0 A	128 fl oz
A	2 lb ai		1 gal

This should be computed as

00.375 lb ai × 1 gal × 2.0 A × 128 fl oz ÷ 1 A ÷ 2 lb ai ÷ 1 gal =

Answer = 48 fl oz of Dimension 2EW

How Much Does This Herbicide Cost Per Acre (or 1,000 ft^2)?

General formula:

Rate	Cost	Conversion factors
Area	Unit size	

Example 6

What is the cost per acre for Trimec 992 when applied at a rate of 4 pints per acre at a price of $83 per 2.5 gallons?

4 pt	$83	1 gal
1 A	2.5 gal	8 pt

This should be computed as

4 pt × $83 × 1 gal ÷ 1 A ÷ 2.5 gal ÷ 8 pt =

Answer = $16.60 per acre

To determine the cost per 1,000 ft^2, simply divide your answer in cost/acre by 43.56 to calculate cost per 1,000 ft^2.

Answer = $0.38 per 1,000 ft^2

Plant Growth Regulator GDD Calculation Math

Given the sample weather data below for a two-week period, you can calculate a growing degree day (GDD) value for each day as well as the accumulated GDDs for the 14-day period. In the table, the temperature values were recorded in Fahrenheit and then converted to Celsius. You should use 0°C as your base temperature in the calculations for cool-season grasses.

Note: In this example, the turf manager's goal was to reapply the PGR after 200 GDDs accumulated, which occurred on day 10, which reset the cumulative GDD.

Day	PGR Application	Low (°F)	High (°F)	Low (°C)	High (°C)	Daily GDDs	Cumulative GDDs
1	First application	48	64	9	18	13	13
2	No	51	69	11	21	16	29
3	No	57	73	14	23	18	47
4	No	62	76	17	24	21	67
5	No	65	81	18	27	23	90
6	No	62	85	17	29	23	113
7	No	68	84	20	29	24	138
8	No	67	88	19	31	25	163
9	No	66	92	19	33	26	189
10	No	70	87	21	31	26	215
11	Second application	68	85	20	29	25	25
12	No	65	88	18	31	25	50
13	No	64	93	18	34	26	76
14	No	67	90	19	32	26	102

First, convert temperatures from degrees Fahrenheit to Celsius by subtracting 32 from the value, and divide that number by 1.8.

Next, calculate the daily GDDs using this formula:

GDD = [(high temperature + low temperature) ÷ 2)] - base temperature

In this example, the base temperature is 0°C, so the average temperature in degrees Celsius was the same as the daily GDD value. For warm-season PGR models, use a base temperature of 10°C in the equation.

Keep track of the running total of GDDs that accumulate, then reapply the PGR when you reach your reapplication interval target (see *Plant Growth Regulator Suppression and Suggested Reapplication Intervals*, page 116).

Reset the model the morning you apply a PGR (as on days 1 and 11 in this example). As a result, the cumulative GDDs on day 1 and 11 were the same as the daily GDDs for those days.

At the end of day 14, 102 GDDs would have accumulated since the last PGR application. A total of 215 GDD would have accumulated between the first and second PGR application.

Additional Resources

Applicator Information

Herbicide labels are available from: CDMS (*https://www.cdms.net/labelssds/home/*), TIMRS Network (*https://www.tirmsdev.com/*), and manufacturer websites.

The Office of Indiana State Chemist provides a Commercial Applicator Pesticide Record Sheet at *https://oisc.purdue.edu/pesticide/commercial_app_use_records.html.*

Turf Programs

Various turf extension publications, newsletters, and information about upcoming educational events are available from the authors and their institutions. Visit the websites to access this information and learn about turf degree programs and career opportunities.

State	University	Turf Program Website
Illinois	University of Illinois	n/a
Indiana	Purdue University	*https://turf.purdue.edu/*
Iowa	Iowa State University	*https://www.extension.iastate.edu/turfgrass/*
Kansas	Kansas State University	*https://www.k-state.edu/turf/*
Kentucky	University of Kentucky	*https://ukturf.ca.uky.edu/*
Maryland	University of Maryland	*https://extension.umd.edu/resources/yard-garden/lawns/lawn-care-and-maintenance/*
Michigan	Michigan State University	*https://www.canr.msu.edu/turf/*
Minnesota	University of Minnesota Twin Cities	*https://turf.umn.edu/*
Missouri	University of Missouri	*https://extension.missouri.edu/programs/turfgrass-science*
Nebraska	University of Nebraska-Lincoln	*https://turf.unl.edu/*
New Jersey	Rutgers University	*https://turf.rutgers.edu/*
New York	Cornell University	*https://turf.cals.cornell.edu/*
North Dakota	North Dakota State University	n/a
Ohio	The Ohio State University	*https://buckeyeturf.osu.edu/*
Pennsylvania	Penn State University	*https://plantscience.psu.edu/research/centers/turf*
Wisconsin	University of Wisconsin-Madison	*https://turf.wisc.edu/*

Diagnostic Laboratories

To identify weeds, turf species, and turf problems, submit samples to the various plant and pest diagnostic laboratories in the region. For details about sample submission, visit the laboratory websites listed here.

State	Diagnostic Laboratory Name	Website	Phone
Illinois	University of Illinois Plant Clinic	*https://extension.illinois.edu/plant-clinic*	(217) 333-0519
Indiana	Purdue Plant and Pest Diagnostic Laboratory	*https://ag.purdue.edu/department/btny/ppdl/*	(765) 494-7071
Iowa	Iowa State University Plant and Insect Diagnostic Clinic	*https://yardandgarden.extension.iastate.edu/pidc*	(515) 294-0581
Kansas	Kansas State University Plant Disease Diagnostic Laboratory	*https://www.plantpath.k-state.edu/extension/diagnostic-lab*	(785) 532-5810
Kentucky	Plant Disease Diagnostic Laboratories	*https://plantpathology.ca.uky.edu/extension/diagnostic-laboratories*	(859) 257-8949
Maryland	University of Maryland Plant Diagnostic Lab	*https://extension.umd.edu/programs/agriculture-food-systems/program-areas/fruit-vegetable-production/plant-diagnostic-laboratory/*	(301) 405-1611
Michigan	Michigan State University Diagnostic Services	*https://www.canr.msu.edu/pestid/*	(517) 355-4536
Minnesota	University of Minnesota Plant Disease Clinic	*https://pdc.umn.edu/*	(612) 625-1275
Missouri	University of Missouri Plant Diagnostic Clinic	*https://extension.missouri.edu/programs/plant-diagnostic-clinic*	(573) 882-3019
Nebraska	University of Nebraska-Lincoln Plant and Pest Diagnostic Clinic	*https://plantpathology.unl.edu/extension-outreach/plant-pest-diagnostic-clinic/*	(402) 472-2559
New Jersey	Rutgers Plant Diagnostic Laboratory and Nematode Detection Service	*https://njaes.rutgers.edu/plant-diagnostic-lab/*	(732) 932-9140
New York	Cornell University Plant Disease Diagnostic Clinic	*https://plantclinic.cornell.edu/*	(607) 255-7850
North Dakota	NDSU Plant Diagnostic Lab	*https://www.ndsu.edu/agriculture/academics/academic-units/plant-pathology/plant-diagnostic-lab*	(701) 231-7854
Ohio	C. Wayne Ellett Plant Pest Diagnostic Clinic	*https://ppdc.osu.edu/*	(614) 292-5006
Wisconsin	University of Wisconsin Turfgrass Diagnostic Lab	*https://tdl.wisc.edu/*	(608) 845-2535

Where to Find Additional Resources

Free and for-sale university extension publications are available from each state. For a list of turf-related publications, call or visit the websites below.

State	Store Name	Website	Phone
Illinois	University of Illinois PubsPlus	*https://pubsplus.illinois.edu/*	(217) 333-2007
Indiana	The Education Store	*https://edustore.purdue.edu/*	(765) 494-6794
Iowa	Extension Store	*https://store.extension.iastate.edu/*	(515) 294-5247
Kansas	K-State Research and Extension Bookstore	*https://bookstore.ksre.ksu.edu/*	(785) 532-5830
Maryland	University of Maryland Bookstore	*https://umcp.bncollege.com/*	(301) 314-2665
Michigan	MSU Extension Bookstore	*https://shop.msu.edu/*	(517) 432-1859
Minnesota	University of Minnesota Bookstore	*https://bookstores.umn.edu/*	(612) 625-6000
Missouri	Extension Publications	*https://extension.missouri.edu/publications*	(573) 882-7216
Nebraska	Nebraska Extension UNL Marketplace Store	*https://marketplace.unl.edu/extension*	(402) 472-9053
New Jersey	New Jersey Agricultural Experiment Station	*https://njaes.rutgers.edu/pubs/subcategory.php?cat=5&sub=39*	n/a
New York	The Cornell Store	*https://www.cornellstore.com/books/cornell-cooperative-ext-pmep-guidelines*	(800) 624-4080
North Dakota	NDSU Extension	*https://www.ndsu.edu/agriculture/extension/publication-search?site_search_api_fulltext=&noscroll*	(701) 231-8944
Ohio	CFAES Publications	*https://extensionpubs.osu.edu/all-publications/*	(614) 292-6181
Pennsylvania	Penn State Extension	*https://extension.psu.edu/*	(877) 345-0691
Wisconsin	Learning Store	*https://learningstore.extension.wisc.edu/*	(877) 947-7827

Phone App for Turf Diagnostics

The Purdue Turf Doctor app is the latest modern turfgrass management tool produced by the Purdue Turfgrass team.

The app helps homeowners and land managers diagnose and address turfgrass problems caused by a variety of factors including weeds, insects, diseases, nuisance animals, and abiotic stress. Turfgrass management professionals and garden center personnel can also use the app to improve communication with their customers and build customer confidence.

The flexible yet powerful user interface allows you to

- Quickly access the latest science-based recommendations from Purdue University experts about how to manage more than 160 turfgrass disorders.
- Identify turfgrass disorders by matching patterns, symptoms, and signs to hundreds of high-resolution photos.
- Check diagnoses with detailed descriptions of damage and stages of problem development linked to each photo.
- Search for information about specific disorders by common or scientific name.
- Use integrated filters to refine diagnoses by problem type, time of year, turfgrass species, field pattern, symptoms, signs, weed type, flower color, and more.
- Easily create and add problems to your favorites list for quick reference later.
- Automatically view a list of your most recent searches.

Purdue Turf Doctor is available from the App Store for iOS devices and Google Play for Android devices. For more information, visit https://www.entm.purdue.edu/turfdoctor/

Download on the App Store

GET IT ON Google Play

@TurfDoctorApp

Additional Resources For Weed Control in Turfgrass Species in Southern States

Weed Control Guidelines for Mississippi, Mississippi State University Extension Publication P1532 (covers weed control information for St. Augustinegrass and centipedegrass and other turfgrasses): https://extension.msstate.edu/sites/default/files/publications/publications/P1532_web.pdf.

Turfgrass Pest Control Recommendations for Professionals, University of Georgia Extension publication B 984 (covers weed control information for seashore paspalum): https://extension.uga.edu/publications/detail.html?number=B984&title=turfgrass-pest-control-recommendations-for-professionals.

Pesticide Emergency and Poison Control Centers

Nationwide Phone Numbers

Poison Center: (800) 222-1222 (*This number will automatically connect you to the poison center nearest you.)*
National Pesticide Information Retrieval System (NPIRS): (765) 494-4576
National Pesticide Information Center: (800) 858-7378
CHEMTREC: (800) 262-8200

Regional Phone Numbers

Environmental Protection Agency Region 2 *(NJ, NY, Puerto Rico, and the US Virgin Islands)*: (877) 251-4575
Environmental Protection Agency Region 3 *(DE, DC, MD, PA, VA, and WV)*: (800) 438-2474
Environmental Protection Agency Region 4 *(AL, FL, GA, KY, MS, NC, SC, and TN)*: (800) 241-1754
Environmental Protection Agency Region 5 *(IL, IN, MI, MN, OH, and WI)*: (800) 621-8431
Environmental Protection Agency Region 7 *(IA, KS, MO, and NE)*: (800) 223-0425
Environmental Protection Agency Region 8 *(CO, MT, ND, SD, UT, WY)*: (800) 227-8917

Indiana Phone Numbers

Indiana Department of Environmental Management: (888) 233-7745, *Pesticide Spill Reporting*
Purdue Pesticide Programs: (765) 494-4566, *General Information*
Office of Indiana State Chemist: (765) 494-1492, *Pesticide Certification and Training*
Environmental Protection Agency Region 5: (800) 621-8431

Other State Pesticide Regulatory Agencies

For more information about the administration and enforcement of state pesticide laws, and pesticide applicator certification, licensing, and training, call or visit the websites below.

State	Agency Name	Website	Phone
Illinois	Illinois Department of Agriculture	https://agr.illinois.gov/	(217) 782-2172
Iowa	Iowa Department of Agriculture and Land Stewardship	https://iowaagriculture.gov/	(515) 281-8591
Kansas	Kansas Department of Agriculture	https://www.agriculture.ks.gov/divisions-programs/pesticide-and-fertilizer	(785) 564-6700
Kentucky	Kentucky Department of Agriculture	https://www.kyagr.com/consumer/agricultural-branch.html	(502) 573-0282
Maryland	Maryland Department of Agriculture	https://mda.maryland.gov/plants-pests/pages/pesticide_regulation.aspx	(410) 841-5710
Michigan	Michigan Department of Agriculture and Rural Development	https://www.michigan.gov/mdard	(800) 292-3939
Minnesota	Minnesota Department of Agriculture (MDA)	https://www.mda.state.mn.us/	(651) 201-6000
Missouri	Missouri Department of Agriculture	https://agriculture.mo.gov/plants/pesticides/	(573) 751-5504
Nebraska	Nebraska Department of Agriculture	https://nda.nebraska.gov/pesticide/cert.html	(402) 471-2341
New Jersey	New Jersey Department of Environmental Protection	https://www.nj.gov/dep/enforcement/pcp/bpo.htm	(609) 984-6507
New York	Department of Environmental Conservation	https://dec.ny.gov/regulatory/permits-licenses/chemical-petroleum-lng-pesticide-permits/pesticide-applicator-technician-guidance/certification	(518) 402-8748
North Dakota	North Dakota Department of Agriculture	https://www.ndda.nd.gov/divisions/pesticide-fertilizer-division/pesticide-program	(701) 328-2231
Ohio	Ohio Department of Agriculture	https://agri.ohio.gov/	(614) 728-6201
Pennsylvania	Pennsylvania Department of Agriculture	https://www.pa.gov/agencies/pda.html	(717) 787-4737
Wisconsin	Department of Agriculture, Trade and Consumer Protection	https://datcp.wi.gov/Pages/Programs_Services/PesticidesFertilizersCertificationLicensing.aspx	(608) 262-7588

About the Authors

Aaron J. Patton is a professor at Purdue University, where he serves as Indiana's turfgrass extension specialist. His research program explores weed ecology and control, low-input grasses, and agronomic practices in turfgrass systems. He mentors undergraduate and graduate students, and teaches the senior-level turfgrass science course.

Matthew T. Elmore is an associate extension specialist at Rutgers University. His research program focuses on understanding best management practices and their integration to control weeds in turfgrass systems. His research supports efforts to educate turfgrass industry professionals as well as undergraduate students in the courses he teaches at Rutgers.

Contributors

Cornell University
Jenny Kao-Kniffin, PhD

University of Illinois
Bruce Branham, PhD

Iowa State University
Quincy Law, PhD
Adam Thoms, PhD

Kansas State University
Jack Fry, PhD
Steve Keeley, PhD

Kentucky
Kenneth Clayton

Michigan State University
Thomas A. Nikolai, PhD

University of Minnesota
Jon Trappe, PhD

University of Missouri
Xi Xiong, PhD

University of Nebraska–Lincoln
Amanda Folck, MS

University of Maryland
Mark Carroll, PhD

North Dakota State University
Deying Li, PhD

Ohio State University
David Gardner, PhD

Pennsylvania State University
John Kaminski, PhD

University of Wisconsin
Paul Koch, PhD
Doug Soldat, PhD

www.ingramcontent.com/pod-product-compliance
Lightning Source LLC
Chambersburg PA
CBHW052256270426
44010CB00018BA/43
* 9 7 8 1 6 2 6 7 1 2 2 9 4 *